What people are saying ab⸻

Against the Web

T0019895

Michael Brooks offers us a polemic focused on the battle for ideas. This book is not a discussion of this or that issue but of worldview and narrative. Michael rips away the sophistry inherent in the prevailing right-wing narratives but then offers a humble, explicitly leftist alternative framework. In offering a left framework he also does battle with some of the key toxicities within the US Left that undermine our collective efforts to build a movement for fundamental social transformation. This book left me thinking; which is what I have come to expect from Michael Brooks.

Bill Fletcher, Jr, editor of globalafricanworker.com, former president of TransAfrica Forum, author and activist

Michael called out the nonsense of the so-called Intellectual Dark Web well before anyone else caught onto their cynical games. A brilliant critique of the Right with very sharp insight on some of the shortcomings of the Left, this book is a must-read for anyone looking to understand how dishonest actors spread their propaganda.

Ana Kasparian, host and executive producer of The Young Turks

You don't know it yet, but this is the book you've been waiting for. Reading Michael Brooks' devastating and insightful account of the IDW feels like a breath of fresh air. He meticulously and expertly challenges the shallow platitudes and certainties of a certain cohort of "public intellectuals": by pulling away the curtain of "logic and reason" behind which these men (and yes, they are all men) hide their juvenile arguments. Yet Brooks does not just rebut and break down—he offers a humane and compassionate counterargument. He takes seriously the IDW's vast and hungry audience and suggests that left thinkers offer them camaraderie and not just self-delusion and justification. He also takes the left to task for the way they react to these provocations, but he maintains a sense of proportionality. He does not equivocate or engage in "both sideism." This is a must-read for anyone who has been puzzled by the appeal of the IDW or anyone seduced by their ideas.

Mehrsa Baradaran, author of *The Color of Money: Black Banks and the Racial Wealth Gap*

Michael Brooks has distinguished himself as one of the most intellectually honest and insightful commentators of the populist left. In this essential book, he takes the best arguments of the Intellectual Dark Web, puts them in historical context, and clinically breaks them apart piece by piece. If you want to understand where American and world politics may be headed, you need to understand the online right. And if you want to understand the online right, you need to read *Against the Web*.

Krystal Ball, co-host of The Hill's Rising

Against the Web is a welcome widening of Michael Brooks' perceptive and incisive work. Easily one of the smartest and most informed commentators on his daily web show and podcast TMBS, he has now expanded, and deepened, his political analysis to book form. With his rare internationalist perspective and passionate commentary, this book provides a crucial guide to the prevailing political landscape of our time. Both the worst temptations of the online IDW pseudo arguments, while also making smart critiques of his own "side," this book is original and needed.

Glenn Greenwald, The Intercept

Against the Web

A Cosmopolitan Answer to the
New Right

Against the Web

A Cosmopolitan Answer to the
New Right

Michael Brooks

Winchester, UK
Washington, USA

JOHN HUNT PUBLISHING

First published by Zero Books, 2020
Zero Books is an imprint of John Hunt Publishing Ltd., No. 3 East St., Alresford,
Hampshire SO24 9EE, UK
office@jhpbooks.com
www.johnhuntpublishing.com
www.zero-books.net

For distributor details and how to order please visit the 'Ordering' section on our website.

Text copyright: Michael Brooks 2019

ISBN: 978 1 78904 230 6
978 1 78904 231 3 (ebook)
Library of Congress Control Number: 2020930635

A CIP catalogue record for this book is available from the British Library.

Design: Stuart Davies

UK: Printed and bound by CPI Group (UK) Ltd, Croydon, CR0 4YY
US: Printed and bound by Thomson-Shore, 7300 West Joy Road, Dexter, MI 48130

We operate a distinctive and ethical publishing philosophy in
all areas of our business, from our global network of authors to
production and worldwide distribution.

Contents

Chapter One

Meet the New Right: The Intellectual Dark Web and Capital's Contradictions

Everyone is preoccupied by how the online world is shaping politics. The left and many liberals have been deeply concerned with the right's fluency on platforms ranging from YouTube to Instagram to Twitter, and their ability to use these platforms to push their messages and create an overall political narrative. With authoritarian right-wing governments holding power from the United States to Brazil and Hungary to India, the need to understand and overcome these forces is urgent. This book focuses on the Intellectual Dark Web (IDW), a group exercise in collective self-branding that may already be by the wayside. However, the tactics, ideologies, and arguments used by this group remain relevant for understanding the broader center-right and right-wing ecosystem, and the absolutely necessary changes that the left must make to tell its own more appealing and dynamic story.

The IDW is a group of men that Bari Weiss introduced to the world in a 2018 *New York Times* profile titled "Meet the Renegades of the Intellectual Dark Web." According to Weiss the IDW was a group of maverick intellectuals who, feeling locked out by a relatively new and culturally dominant "political correctness," came together to speak truth to the power of the liberal consensus. According to Weiss, the group was quickly taken up by a public hungry for free thinking, and it is certainly true that the two most prominent members, Sam Harris and Jordan Peterson, were filling auditoriums with admiring fans. By the end of the year, when Amelia Lester called the online magazine Quillette "The Voice of the Intellectual Dark Web" in Politico, everyone likely to read such an article was well familiar with the IDW.

This is how Weiss introduced the IDW in her original piece:

Here are some things that you will hear when you sit down to dinner with the vanguard of the Intellectual Dark Web: There are fundamental biological differences between men and women. Free speech is under siege. Identity politics is a toxic ideology that is tearing American society apart. And we're in a dangerous place if these ideas are considered "dark."

Showing a stunning lack of historical awareness—and by the way, the IDW's stunning lack of historical awareness will be one of the major themes of this book—the subjects of the profile informed Weiss that "a decade ago...when Donald Trump was hosting 'The Apprentice,' none of these observations would have been considered taboo." In reality, both the group's claim to be a persecuted minority and their depiction of the left as censorious and dominant were hardly new accusations. The conservative framing of American politics around a perceived culture war dates back to at least 1951 when *National Review* founder William F. Buckley, who was in that moment both a segregationist and a vocal white supremacist, released his book *God and Man at Yale*. Though the culture-war specifics might not have been firmly in place in that book, they certainly were by the time conservative philosopher Allan Bloom wrote *The Closing of the American Mind* in 1987. When the movie PCU (starring a bald Jeremy Piven) came out in 1994—10 years before the first season of *The Apprentice* and a full 24 years before Bari Weiss's piece hit the *New York Times*—these complaints were shop-worn clichés.

So is the IDW just a rebranding of old-style cultural conservativism? Not exactly, although you might be forgiven for thinking so when you notice that Ben Shapiro is an IDW member in good standing. Shapiro is a religious conservative who believes that Palestinian rights can be disregarded because, as he says in one YouTube clip, "God gave Israel to the Jewish people." (In the video, entitled "Ben Shapiro: Why Jews Vote Leftist," a young Shapiro expresses amazement and disgust that most American Jews don't share this belief.) Like any good fundamentalist, Shapiro is firmly opposed to letting women control their own bodies. He invariably refers to abortion as "killing babies." He regularly speaks out against "open borders," gun control, socialism, and even redistributive taxation. In 2003, 2 years after a teenaged Shapiro began writing a nationally syndicated column (the conservative obsession with teen "prodigies" never ceases to amaze), he used it to cheer on the invasion of Iraq. Shapiro could be grouped together with Rush Limbaugh and Sean Hannity as naturally as he is with his IDW comrades-in-arms Sam Harris and Jordan Peterson.

It's probably Harris, who genuinely does part ways with the Limbaughs and Hannities of the world on a number of core issues, who marks the difference between the IDW and the more old-fashioned right. The Stanford- and UCLA-educated neuroscientist is a warmonger and an apologist for the status quo in many ways I'll explore as the book goes on, but he has conventionally liberal views on domestic policy issues ranging from abortion to closing the gun show loophole. He supported Hillary Clinton against Donald Trump in the 2016 election. And where Ben Shapiro naively believes that God Himself shares his attitudes toward women and Palestinians, Harris is fiercely secular. Long before there was an Intellectual Dark Web, Harris belonged to a group of intellectuals who collectively branded themselves The New Atheists.

While many of my major intellectual influences are in fact atheists of the old school materialist tradition who analyzed religion as a cultural force determined by economics and social relations, I was always critical of the obsessive view of atheism as an innately liberating belief system that superseded the material conditions that we all live in and that shape our lives. The New Atheists, and Harris in particular, spent a lot of time obsessing over the problem that people believe "bad things" even as they ignored the real-world forces that might generate those bad beliefs, and in turn, adopted much of the reactionary worldview of their Christian counterparts in the Bush administration, but we will explore Harris' fixation on "bad ideas" later in the book. For now it's suffice to say that his prime intellectual contribution to New Atheism was to put a scholarly sheen on the belligerent, hysterical, and ultimately imperial neoconservative foreign policy agenda that defined the American right's worldview in the Bush era. Seen from this perspective, his current chummy collaboration with Shapiro is not as surprising as it might otherwise seem.

Still, this move to the IDW milieu certainly represents a step down from the New Atheist scene. Christopher Hitchens was a witty and insightful writer whose post-9/11 turn to the right was preceded by a long and honorable history on the left. Richard Dawkins is not *just* a schmuck on Twitter; he's also a real scientist and a gifted popularizer of evolutionary biology. Daniel Dennett was writing serious academic philosophy long before he started writing for a popular audience.

Compare Hitchens, Dawkins, and Dennett to Harris' new club, which includes failed stand-up comic Dave Rubin as a charter member. In the original *New York Times* piece, Weiss credulously quotes Rubin when he called himself and the rest of the IDW "just a crew of people trying to have the kind of important conversations that the mainstream won't." (If you watch my show, you are undoubtedly reading that quote in the "Rubin voice," which should make the experience much more satisfying.)

Now, Rubin and I exist in the same media ecosystem, hosting YouTube shows and podcasts. (In fairness, his show has a larger audience than mine while my show is the infinitely superior program.) I'm the host of *TMBS* (The Michael Brooks Show) and the co-host of *The Majority Report*, which is part of the TYT (The Young Turks) network, as was Rubin's show until he dramatically "left the left" in 2015. My good friend and frequent collaborator Ana Kasparian knew Rubin during his TYT years. The way she tells it, he "left the left" at least as much as a cynical career move as a genuine ideological shift. I believe her. Even if you think his turn toward "classical liberalism" (read: half-baked libertarianism) was completely sincere, though, here's what anyone who watches his show can confirm for themselves: Dave Rubin doesn't belong in the "intellectual" *anything*. He's dumb as a rock. He might as well *be* a rock to judge by how little he bothers challenging the right-wing guests he "has important conversations" with on his show. He talks a lot about having "high-level" conversations about "ideas," but in practice he stares blankly into space while a parade of crackpots and crypto- and not-so-crypto fascists make ridiculous assertions. His idea of having "important conversations" certainly doesn't include talking to anyone who would seriously challenge *him*. He's been dodging debate challenges from *The Majority Report*'s Sam Seder for years.

To get a sense of why he's so afraid of Seder, check out how he did in what was supposed to be a friendly chat with an ideological ally, the amiable and IDW-aligned Joe Rogan. On episode 1131 of *The Joe Rogan Experience*, Rubin rants about the evils of government regulation. He and Rogan start out by agreeing that bakers shouldn't have to make cakes for gay people's weddings— gliding over the possible civil-rights implications for legislation all over the country. Rubin, as usual, gets a lot of mileage out of *being a married gay man himself*. The two further agree that left-wing objections to inherited wealth are misguided. But when Rubin says that the government doesn't do *anything* right, Joe Rogan reaches the point where he can't go along with Rubin's increasingly absurd assertions.

Rubin: Do they do the Post Office well? No! What do they do well?

Rogan: They do the Post Office pretty good, actually.

Rubin: But guess what, if the Post Office closed tomorrow, it would be all right. You'd still get mail. Amazon would—

Rogan (drily): It would suck.

Rubin: No it wouldn't. Amazon would pick—

Rogan: You'd have to send things through UPS, it would cost a lot more...

Rubin: It wouldn't, though. Competition would start kicking in and between UPS, Fed Ex and Amazon and drones and blah blah blah...

Rogan doesn't even make the obvious points about how much Amazon currently relies on the Post Office or how it would be massively unprofitable for private companies to service depopulated rural areas in a post-USPS world without enormously jacking up prices—all excellent reasons to think that it would indeed suck a great deal—but I have a hard time imagining that anyone watching the exchange or listening to it later could have missed the way that Rogan is bringing up practical realities while Rubin is both literally and figuratively hand-waving it all away. It gets even funnier when Rubin tries to back up his childlike belief in the invisible hand by telling a story about ordering live chickens from UPS. Rogan points out that Rubin's story is actually a USPS success story. UPS doesn't deliver live chickens. Flustered, Rubin concedes that his story *was* about the USPS, but insists that in a libertarian utopia UPS *would* deliver live chickens and that they'd do it even better. He then awkwardly pivots from the postal service to regulation.

Rubin: I'm not saying these things have to be eliminated tomorrow, I'm not even *really* calling for them to be eliminated, but just generally, what problem would you—

everything you're building here right now...do you want the government to tell you how to do all these things, and all the regulations that you gotta have your electric wire like this and...

Rogan (slowly): Regulations like that for construction are important, though. You got to make sure that people don't do stupid shit, that you don't have power lines near a water line, and that...

Rubin: But I would put most of that on the builders, though. They want to build things that are good.

Rogan (after making incredulous noises): That's not true. People cut corners *all the time*. You have to have regulation when it comes to construction methods or people are going to get fucked.

Rubin: They cut corners when there are regulations anyway.

Rogan (patiently): They do, but they would do a *lot* more if there weren't regulations. You go to Third World countries and look at construction methods, they're fucking dangerous. That's why schools collapse on kids...

At this point, Rubin backs off again, retreating to an even vaguer version of some undefined libertarian claim about regulation—a claim that he says he likes "intellectually," whatever that means. (I strongly suspect that it's his way of acknowledging that it's impractical bullshit.)

Rubin will appear from time to time in this book either as an illustration of the bankruptcy of some of the IDW's ideas or as comic relief. But we shouldn't get too fixated on him. In this case, to be fair to Rubin, he's making exactly the same argument that Milton Friedman made for decades. In both the "smart" and "dumb" (Rubin) versions, the claims are equally untenable. This is also a good illustration of why, if we are going to—as the IDW people obsessively say but seldom do—"steelman" our opponents' arguments, I will spend less time focusing on

Rubin and his various antics than the comedic part of me, and probably many of you, so desperately wants me to do. But don't worry—there will be some of that. The centrality of Rubin to the IDW's project undermines its claims to analytic rigor and a culture of intellectual introspection, which is probably why the IDW-aligned website *Quillette* has spent so much time lately isolating Rubin and his silly show. (Rubin has taken this about as well as you'd think.)

It would be easy to spend the book mocking the idiocy of people like Dave Rubin (which admittedly is a core brand proposition for me) and ripping into the hypocrisy of guys like Jordan Peterson, who talk as if they're being persecuted by the all-powerful Intolerant Left while they hold down a tenured position at a prestigious university (Peterson), host a super-popular podcast (Harris), write best-selling books (Peterson, Harris, Shapiro), and receive loving profiles in mainstream media outlets (all of them). And don't get me wrong, I *will* do some of that—but I'm primarily interested in a broader and more important project.

For one thing, I want to understand not just what the IDW has in common with previous groups of reactionaries, but what's new and different about it, since even after the "Intellectual Dark Web" withers away, the new right will continue in the same vein. It will, for instance, continue to hide its conservatism. That Harris is an atheist with some socially liberal domestic policy positions, that Rubin is a married gay man, that even Jordan Peterson never *quite* calls himself a conservative—all of this helps them brand themselves as unclassifiable renegades even as they share elements of an unmistakable anti-left agenda.

They all defend the capitalist economic order domestically and American imperial hegemony globally. They all see themselves as defenders of a poorly understood (and frankly historically illiterate) construct called "the West." They all defend what they imagine to be "biology" against feminists, and at least some of them—like Sam Harris, who's supported the odiously far-right and overly bigoted Charles Murray—defend a similar stance when it comes to race. Crucially, in all of these areas the IDW promotes narratives that either *naturalize* or *mythologize* historically contingent power relations—between workers and bosses, between men and women, they are old school reactionaries.

But, how is the IDW different than what came before?

That's a bit more complicated, a bit more difficult to explain. I think that the primary difference isn't to be found in the IDW itself, but in the larger context, the historical moment, that they arose within.

To understand the IDW moment we should look back to April 1917, when Vladimir Lenin returned to Russia from his exile in Switzerland. The train station at which he arrived has long been a symbol of the revolution he went on to lead. But, 100 years later, in his contribution to a series of articles the *New York Times* published to mark the anniversary of that revolution, *Jacobin* editor Bhaskar Sunkara contrasted the political possibility that continues to be represented by St Petersburg's Finland Station with two contemporary metaphorical alternatives.

"Singapore Station" is the unacknowledged destination of the neoliberal center's train. It's a place where people in all their creeds and colors are respected — so long as they know their place. After all, people are crass and irrational, incapable of governing. Leave running Singapore Station to the experts..."Budapest Station," named after the powerful right-wing parties that dominate Hungary today, is the final stop for the populist right. Budapest allows us to at least feel like we're back in charge. We get there by decoupling some of the cars hurtling us forward and slowly reversing. We're all in this together, unless you're an outsider who doesn't have a ticket, and then tough luck.

Sunkara is no uncritical apologist for everything that happened in the years after Lenin arrived at the Finland Station. In his article, he emphasizes "political pluralism, dissent, and diversity" as integral parts of his socialist project. He's dead right, though, that the pressing threats to democracy in today's world come from the "decidedly non-Stalinist" forms of authoritarianism represented by Singapore and Budapest Stations.

Like Bhaskar, I believe in Finland Station. I'm also aware that much of the world today does often seem to be pitted between Budapest and Singapore. The 2016 Presidential election in the United States, the struggles within Britain's political elites over Brexit, and the corporate media faith in modern "Centrist" leaders like Emmanuel Macron all manifest a struggle between these two options. That said, if we see them as irreconcilably opposed, we'll miss important parts of the larger picture. Go back in the archive and look at Tony Blair's Home Secretary David Blunkett's comments on migrants to see that the themes of today's rightists did not emerge in a magical vacuum. Even if representatives of Budapest sometimes express dissatisfaction with market fundamentalism, the two have more in common than simplistic media narratives reveal. Look no further than the relationship between the "woke" Justin Trudeau government in Canada and the decidedly unwoke government of Brazil's Jair Bolsonaro. In particular, these two agendas have found ways to not only co-exist but cross-pollinate in emergent forms of right-wing politics—particularly in new and social media.

It's with this context in mind that I'll be exploring the IDW. I'll treat them as a case study in the way that reactionaries have begun to repackage their project of defending traditional hierarchies even as I try to show what a rejuvenated, humane, internationalist, and appealing version of the politics of the Finland Station might look like.

Along the way I'll show exactly what's wrong with the IDW's arguments. Largely ignoring the always easy to dunk on dummy

Dave Rubin and/or the bland Weinstein brothers, I'll take aim at the misleading narratives of Shapiro, Harris, and Peterson. Where they naturalize or mythologize social problems, I am going to *historicize* them. I'll critique the inadequate ways that the left has responded to the IDW's challenge and the broader evolution of right-wing ideology that its members represent. Though I firmly reject false equivalencies between well-intentioned but misguided leftists and actively malevolent reactionaries, I don't ignore the way that counterproductive strategies and inadequate analyses have played into the enemy's hands. Finally, I'll sketch out a left-wing vision that might help us meet the challenges presented by the new right by providing those who currently listen to Harris, Peterson, and Shapiro with a better way of understanding the world. Saying that culture-war skirmishes have the effect of distracting us from the economic forces that lie at the root of our problems is true enough, but it's also not enough. The mistakes, excesses, and wrongs of the performatively ultra-woke can't be combated with economic analysis alone. Culture matters. In Marxist terminology, the ideological and cultural superstructure rests on a material base, but that doesn't mean that the latter issues don't have a life—and an impact—of their own, or that we're going to win power by telling people to simply ignore the cultural issues that profoundly impact their lives. What we need to transcend the stale dichotomies of the past is a cosmopolitan vision of global socialist humanism.

In writing this book I took particular inspiration from the Nobel-Prize winning economist Amartya Sen, who, in response to the usual Eurocentric claim that the tradition that produced human rights flowed from Greece through Rome through Europe to America, countered that rather than a story of cultural continuity, the struggle for human dignity has always been fragmented, multicultural, and global. The West's history is hardly an unbroken chain of progress toward social equality.

11

Instead of inaccurately particularizing the concepts of rights and justice as Western, we should understand that the fight for social equality and justice has arisen, in various forms, in a variety of cultures from ancient to modern times.

While Jordan Peterson talks about "the West" as having discovered individual rights in a way that's so ahistorical that to listen to him you'd think that the UN Declaration of Human Rights was extracted from a speech Achilles gave at the end of the *Iliad*, the cynical promoters of "Asian values" are the other side of the same coin. Putting a minus where the Petersons of the world put a plus, leaders in Singapore and Malaysia have argued that their disregard for individual "liberal" protections—that any socialist must defend—of assembly and free speech and dissent is grounded in Confucian group norms. Sen elegantly and amply demonstrates that there are Asian, African, and Islamic arguments for open societies and free debate. Both the European chauvinist's narrow and bigoted claim that "West is the Best" and the despotic case for "Asian" values should be rejected—not because of some moralistic "taboo," but because the historical narrative underlying both arguments is a patchwork of nonsense.

Similarly, we should reject *both* the Shapiro/Peterson defense of traditional hierarchies and the misguided attempts of the ultra-woke to improve society by scolding people for holding imperfect ideas in their heads (or for having senses of humor). Wittingly or unwittingly, ultra-woke scolds feed a project of endless fragmentation and standpoint epistemology that, if relied on as a strategy for action, destroys any possibility for collective liberatory endeavor. At the same time, we need a path forward that rejects empirically baseless racial essentialism while avoiding the descent into tone-deaf economic reductionism. This is the only way we will move toward an equitable, compassionate, and truly global socialist future.

The following framework and synthesis, which borrows from

Marx, Fatima Mernissi, Cornel West, Adolph Reed, Bill Fletcher Jr., Mehrsa Baradaran, and many others, begins by grounding the critique of the IDW, the right, and capitalism in material conditions, as understanding these realities is essential to the success of any left project. It then elucidates a liberatory and Internationalist project that has broad cultural appeal and is rooted in an ethos of openness and dynamism, not puritanical moralism.

The Cape Verdean theorist and revolutionary Amilcar Cabral noted that imperialism and colonialism pushed its subjects outside of history, and that the purpose of anti-colonialism was to return the history-making process to colonized peoples. This book is best understood as an act of *historicizing* that integrates an international socialist project to both counter the right's fixation on pop science, hierarchy, and mythology and, ultimately, to build a better world.

However, only a bit more than a year into this joint project, there are already significant fissures inside this aggrieved band of renegades. The two leading lights of the IDW, along with Ben Shapiro and the Weinstein brothers, have left their mark on pop-intellectual culture in the United States and Europe. The way they combined and repackaged the agendas of Budapest and Singapore isn't going anywhere. The group's devotion to affirming capitalism when its legitimacy is under threat, its shared obsession with campus and social media controversies— as we'll see, they manage to get such controversies wrong even when they're right—and their intense interest in IQ and other innate justifications for systemic inequalities is the defining feature of the right-wing project historically and in our specific moment.

Chapter Two

"The History is Completely Irrelevant"

I've said that the Intellectual Dark Web's favorite move is to naturalize or mythologize historically contingent power structures. How this works is pretty obvious in the case of an ignorant charlatan like Jordan Peterson. In the course of defending hierarchies, Peterson merrily bounces back and forth between calling himself an "evolutionary biologist" (a field he has no academic background in whatsoever) to elucidating on "the dragon of chaos." Sam Harris, though, is a different beast requiring more serious investigation.

Harris first became a public figure with the release of his anti-theistic manifesto *The End of Faith* in 2004. Six years after declaring that religious belief is dangerously irrational and that politely pretending otherwise is "a luxury that we can no longer afford," Harris wrote *The Moral Landscape*. The book dismissed any attempt to build a system of moral values on any foundation other than the empirical sciences as nonsense. The next year, he followed this up with a thin book titled *Free Will* in which he affirmed that:

> Our wills are simply not of our own making. Thoughts and intentions emerge from background causes of which we are unaware and over which we exert no conscious control. We do not have the freedom we think we have.

All in all, the worldview of Harris' books might seem like a far cry from the one promoted by the Canadian academic Jordan Peterson. Where Peterson is as emotionally intense as a tent revival preacher, Harris' speech patterns are usually calm and measured. The impression he works hard to convey is that of a rational man inviting you to face harsh, and often unpleasant, realities.

He even manages to sound like that when he's floating the idea that—while it would be a terrible shame, of course—America might *have to* commit genocide in the Middle East. Here's the passage in *The End of Faith* in which he promotes this notion:

It should be of particular concern to us that the beliefs of Muslims pose a special problem for nuclear deterrence. There is little possibility of our having a cold war with an Islamist regime armed with long-range nuclear weapons. A cold war requires that the parties be mutually deterred by the threat of death. Notions of martyrdom and jihad run roughshod over the logic that allowed the United States and the Soviet Union to pass half a century perched, more or less stably, on the brink of Armageddon. What will we do if an Islamist regime, which grows dewy-eyed at the mere mention of paradise, ever acquires long-range nuclear weaponry? If history is any guide, we will not be sure about where the offending warheads are or what their state of readiness is, and so we will be unable to rely on targeted, conventional weapons to destroy them. In such a situation, the only thing likely to ensure our survival may be a nuclear first strike of our own. Needless to say, this would be an unthinkable crime—as it would kill tens of millions of innocent civilians in a single day—but it may be the only course of action available to us, given what Islamists believe. How would such an unconscionable act of self-defense be perceived by the rest of the Muslim world? It would likely be seen as the first incursion of a genocidal crusade. The horrible irony here is that seeing could make it so: this very perception could plunge us into a state of hot war with any Muslim state that had the capacity to pose a nuclear threat of its own.

Even if you've seen that passage before, it's worth taking a moment to allow its full absurdity to register. Harris is both trying to sell his readers on the possible "need" to engage in a genocidal mass killing and preemptively demanding that no one think badly of him for advocating this. After all, he's already said that such a strike would be "horrible," "unconscionable," even an "unthinkable crime."

The phrase "virtue signaling" is wildly overused—including by Harris himself. Reactionaries often refer to any attempt to condemn bigotry as "virtue signaling." But what Harris is doing at the end of that passage is virtue signaling in its purest form. He doesn't want the genocidal *content* of the passage to be held against him. Instead, he wants to be given credit for having noisily signaled his virtue even as he promotes a "necessary" genocide that he, as a "good" person, naturally finds upsetting.

The endless heat generated by this paragraph has resulted from critics of Harris pointing out how horrifying his vision is while his supporters insistently claim that this moral condemnation is somehow unfair. I suppose it would be unfair to lambast Harris if in the preceding sentence he had written that he "would be absolutely hysterical, immoral and ludicrous, to seriously assert the following..." But he didn't. And furthermore, Harris, who often presents himself as the *sine qua non* of rational thought, completely disregards the realities of international arms control as they existed when he wrote *The End of Faith*. When Harris published his book, the Bush administration was waging a global "war on terror" that had already claimed hundreds of thousands of lives; a war that was predicated on the assertion that the United States was justified in taking aggressive preemptive action to prevent weapons of mass destruction from harming US national security. Of course, the only avowedly Islamic state with a nuclear weapon in 2004— and still today—was Pakistan, which had never engaged in a jihadist "dewey-eyed" first strike, even as tensions ratcheted up with India. Harris' "hypothetical" scenario ran roughshod over what was happening in the real world and gave cerebral cover—and moral license—to some of the most dangerous policymakers in Washington. In addition to being morally appalling, Harris' take was historically and politically illiterate.

This is also true of his method. In his subsequent convoluted rationalizations, Harris has referred to this passage as a "thought experiment." It's important to spell out why that's wrong. According to my friend Ben Burgis, the author of *Give Them An Argument: Logic for the Left*, a thought experiment generally refers to two things: first, an imaginary situation designed to test whether a certain definition of a concept captures what we really mean by it, and second, an imaginary situation in which we bring two moral principles into conflict in order to discover which one we care more about. The most famous thought experiment is the so-called "Trolley Problem," which was originally formulated by the British philosopher Philippa Foot, though the version that most people are familiar with incorporates a change suggested by the American philosopher Judith Jarvis Thomson. Here's Foot's original example:

Edward is the driver of a trolley whose breaks have just failed. On the track ahead of him are five people; the banks are so steep that they are not able to get off the track in time. The track has a spur leading off to the right, and Edward can turn the trolley onto it. Unfortunately, there is one person on the right-hand track. Edward can turn the trolley, killing the one; or he can refrain from turning the trolley, killing the five.

Most of us think it's wrong to intentionally kill innocent people. The problem is that most of us *also* think that it's wrong to knowingly let innocents die when we could have saved them. Some moral philosophers, like Immanuel Kant, think the second principle is more important than the first—that *killing* is worse than *letting die*. Other philosophers are 'consequentialists,' meaning that they believe morality is fundamentally about maximizing good consequences. The consequentialist approach to the Trolley Problem is to make whatever decision—in this case, turning the trolley to the right—that will result in the fewest deaths.

When they first hear Foot's version of the Trolley Problem, the majority of people have a consequentialist reaction. (Or their eyes glaze over, as yours might be doing. Just give me a minute here. This is going to come up later.) The usual response is to argue that the morally "right" thing for Edward to do is to turn the trolley to the right, killing the one person to save the five. Nevertheless, Judith Jarvis Thomson amended the Trolley Problem in such a way that, when hearing her version, people have the opposite reaction. Here is Thomson's version of the problem:

> George is on a footbridge over the trolley tracks. He knows trolleys and he can see that the one approaching the bridge is out of control. On the track back of the bridge there are five people; the banks are so steep that they will not be able to get off the track in time. George knows that the only way to stop an out-of-control trolley is to drop a very heavy weight into its path. But the only available, sufficiently heavy weight is a fat man, also watching the trolley from the footbridge. George can shove the fat man onto the track in the path of the trolley, or he can refrain from doing this, letting the five die.

When presented with this version of the Trolley Problem, most people refuse to sacrifice the fat man's life to save the five people. In other words, though on the face of it the moral calculation in both Trolley Problems is the same—in both versions of the story, one person dies to save five—the different responses that people give demonstrate that in real life, people distinguish between actively participating in a killing and letting someone die.

Whatever you think about the *solutions* to the Trolley Problems, you can see the point of the thought experiment. Two principles are being pitted against each other to test which one we think 'outranks' the other. But what's the point of Harris' "thought experiment" on Muslim genocide? What concept is it supposed to clarify? What is it supposed to show us about how we should *think*?

Remember, clarification is the only thing a 'thought' experiment *can* do. If you want to know how the world actually *works*, you need to do an experiment-experiment. Or a historical or sociological study. Or find some other way to go out into the world and *check* to see what's true. Unless Harris wants to start claiming that his meditation practice has given him psychic powers, there's just no way to gain empirical information from his armchair. (Unfortunately, my own meditation practice has yet to give me these powers, though not for lack of trying.)

I support Medicare for All. This goes right to the core of my politics and my values. Imagine that I made a Harris-style argument in favor of it by deploying a "thought experiment."

Private insurance poses serious moral hazards and may in the long run undermine public health, despite the possible gains from private sector innovation. The danger of private insurance is such that—and this is horrible to contemplate— we may need to consider a government solution to healthcare. Now in some major respects this is unthinkable—however, it's something that we need to examine no matter how upsetting.

Would framing my support for Medicare for All in this way make it somehow better than a straightforward argument that emerged from clearly stated premises about moral hazards and public outcomes? Would libertarians who opposed such government programs on principle be mollified by the rhetorical bones I was throwing them by talking about how horrible and unthinkable it was that we might maybe possibly need to institute this particular program? They wouldn't, and, given their worldview, they *shouldn't*. Any libertarian with half a brain would recognize that the important thing is the position I was taking, not how I was dressing it up.

Harris often claims that his remarks are taken out of context, but in this case, a fair reading of the context makes it very clear that Harris' discussion of nuclear genocide is not a thought experiment in any meaningful sense. He's describing something he thinks may actually happen. In fact, as he underlines in the sentences immediately following the ones quoted above, he believes that it is *extremely likely* that millions will be eradicated as the result of a nuclear explosion prosecuted by non-State terrorists.

All of this is perfectly insane, of course: I have just described a plausible scenario in which much of the world's population could be annihilated on account of religious ideas that belong on the same shelf with Batman, the philosopher's stone, and unicorns. That it would be a horrible absurdity for so many of us to die for the sake of myth does not mean, however, that it could not happen. Indeed, given the immunity to all reasonable intrusions that faith enjoys in our discourse, a catastrophe of this sort seems increasingly likely. We must come to terms with the possibility that men who are every bit as zealous to die as the nineteen hijackers may one day get their hands on long-range nuclear weaponry. The Muslim world in particular must anticipate this possibility and find some way to prevent it. Given the steady proliferation of technology, it is safe to say that time is not on our side.

Then there's the larger historical context. *The End of Faith* was published about one year after George W. Bush invaded Iraq in

March 2003. A major part of the pretext for war was the allegation that Iraq had "an active nuclear program" (along with various other "Weapons of Mass Destruction") and that Saddam Hussein might share such "WMDs" with fundamentalist terrorists—in Harris' words, "men who are every bit as zealous to die as the nineteen hijackers."

In the lead-up to the invasion of Iraq, Bush declared that Iran was part of an "Axis of Evil" whose other members were Iraq and North Korea. Iran and the United States have a long and complicated history. In 1953, the United States helped overthrow Mohammad Mosaddegh, Iran's democratically elected leader; thereafter, the country was ruled by the US-backed Shah Mohammad Reza Pahlavi, a brutal dictator. Once the Iranians deposed the Shah in 1979 and replaced him with Ayatollah Khomeini, the United States imposed sanctions of various kinds that continue to this day. In 2006—2 years after the publication of *The End of Faith*—the US and its allies persuaded the UN Security Council to pass Resolutions 1696 and 1737. The former demanded that Iran cease its nuclear enrichment program and the latter imposed new and more punishing sanctions for noncompliance. For its part, Iran insisted that the only aim of its enrichment effort was to develop a civilian nuclear energy program. Nevertheless, the Bush administration and its allies refused to accept Iran's arguments, partially because they, like Harris, believed that the "steady proliferation" of nuclear technology in "the Muslim world" would generate a "scenario in which much of the world's population would be annihilated on account of religious ideas."

It's important to keep in mind that the Bush administration had rejected overtures from a reformist Iranian leadership headed by President Mohammad Khatami, which had not only condemned the 9/11 attacks but also offered Iranian assistance to the United States' efforts in Afghanistan, assistance that was motivated primarily by Iran's own geostrategic interests. Iran, like most countries, acts according to its geostrategic interests, and at the time it believed that it was wise to reach a détente with the United States. In other words, though they no doubt govern a repressive religious state, Iranian leaders—unlike their American counterparts—are not looking to remake, let alone destroy, the world. It is for this reason that Harris' subtle conflation of the actions of an independent terrorist network like al-Qaeda with those of a nation-state like Iran is ludicrous. The most basic historical overview would have shown Harris that many revolutionary states, from Stalin's Soviet Union to Mao's China to the Ayatollah's Iran, have very quickly determined that their geopolitical decisions cannot be based on ideological fervor, but must rather be premised on the cold calculations of *Realpolitik*. This is simple stuff, and it is appalling and embarrassing that Harris' "thought experiment" isn't informed by any knowledge of historical facts.

In fact, if we'd listened to Harris we might very well have had a third war in the greater Middle East. Instead, under the Obama Administration (following an approach led by Brazil's President Lula Da Silva), Iran and the G5+1 concluded a nuclear agreement (until the Trump Administration ripped it up due to its own belligerence and warmed-over neocon tendencies). Harris, unsurprisingly, has maintained a "mindful" silence on most of this.

But perhaps none of this is relevant. In his frequently updated "Response to Controversy" blog post, Harris has claimed that the warning he issued during the Bush era—when he said, remember, that nuclear war was "plausible," that indeed if "the Muslim world" didn't find a way to prevent it, it probably *would* happen sooner or later ("time is not on our side")—doesn't actually have anything to do with the real world.

Clearly, I was describing a case in which a hostile regime that is *avowedly suicidal* acquires *long-range* nuclear weaponry (i.e.

they can hit distant targets like Paris, London, New York, Los Angeles, etc.). Of course, not every Muslim regime would fit this description. For instance, Pakistan already has nuclear weapons, but they have yet to develop long-range rockets, and there is every reason to believe that the people currently in control of these bombs are more pragmatic and less certain of paradise than the Taliban are. The same could be said of Iran, if it acquires nuclear weapons in the near term (though not, perhaps, from the perspective of Israel, for whom any Iranian bomb will pose an existential threat).

The contrast between the 24 episode-level hysterical bloodlust of the passage from *The End of Faith* and this mealy-mouthed revisionism is so stark that Harris' attempt to say that this is "clearly" what he meant can be passed over with the contempt that it deserves. Notice, though, that even here, he's trying to have it both ways. Is the Iranian government "avowedly suicidal" enough to initiate a nuclear exchange with Israel—or are they "more pragmatic and less certain of paradise" than that? (For some reason, he seems to think that Iran would be willing to annihilate itself by starting a war with Israel—a nuclear power— but would not be willing to do so by initiating strikes on "Paris, London, New York, Los Angeles, etc.") Keep in mind that the original passage was about an Islamist "regime" acquiring nuclear weapons. If this was a not-even-very-long-term danger in 2004 (though why say "time is not on our side"?) then which regime *was* he talking about? He mentions the Taliban, but it hadn't held state power for 2 years by the time Harris wrote that passage, and when it *did*, its actions hardly resembled those of a cartoonish nation-state whose government lacked any sense of self-preservation. (In fact, as I'm writing this some factions of the Taliban are engaged in peace negotiations with the United States in Qatar.) But if Harris isn't talking the Taliban, and if he isn't talking about Pakistan, and he maybe even isn't exactly talking

23

about Iran, who exactly is he talking about? I'm pretty sure he wasn't musing about a nuclear first strike coming from America's long-term strategic partner Saudi Arabia. And if the Saudis too are struck off our list of possibilities, we've come pretty close to running out of candidates for the "Islamist regimes" that grow "dewey-eyed at the mere mention of paradise" discussed in *The End of Faith*.

Despite the impossibility of squaring his current rationalizations with the actual words he wrote in 2004, Harris complains in "Response to Controversy" that he's being "defamed."

> Such defamation is made all the easier if one writes and speaks on controversial topics and with a philosopher's penchant for describing the corner cases—the ticking time bomb, the perfect weapon, the magic wand, the mind-reading machine, etc.—in search of conceptual clarity...

Another whimsical place Harris' "philosopher's penchant for describing corner cases" has taken him is the possibility—only under "extreme circumstances," of course—that it might be necessary to torture suspected terrorists. As might be expected, he takes no pleasure in advocating torture. It's just another thing—like committing genocide by nuclear first strike—that the United States might *have* to do.

> I am not alone in thinking that there are potential circumstances in which the use of torture would be ethically justifiable. The liberal Senator Charles Schumer has publicly stated that most US senators would support torture to find out the location of a ticking time bomb. Such scenarios have been widely criticized as unrealistic. But realism is not the point of these thought experiments.

Even "liberal" Chuck Schumer, the senator so progressive that he voted for the invasion of Iraq and whose extreme far-right views on Israel would not be out of place at an APAC convention, in fact they're not as he's a speaker at APAC regularly.

The important thing is that we shouldn't think too hard about the *realism* of the ticking-time bomb scenario. Who, after all, is even talking about the real world? The United States has spent decades fighting pointless wars in the Middle East; Harris is just exercising his philosopher's penchant for abstract thought. Meanwhile, the United States is *actually doing* the things Harris' "thought experiments" just so happen to be about, like torturing suspected terrorists and going to war with the stated goal of stopping Islamic fundamentalists from gaining access to weapons of mass destruction. Somehow, his philosopher's penchant for exploring corner cases never led him to lay out thought experiments in which Iraqis or Iranians or Afghanis or Palestinians were forced by extreme circumstances to fight off occupying powers by using extreme tactics. Such circumstances are far outside the reach of Harris' imagination, empathy, or analysis.

Of course Harris has no critique of the invasion of Iraq, but only faults the Bush administration for being naive about the likelihood of sectarian conflict and violence in an occupied Iraq.

Even in 2015, 12 years into this horror show, the only criticisms Harris could muster of either the invasion or the neocons who oversaw it were that the war was a "risky" decision and that in retrospect Bush and his cronies were a bit too "idealistic" and impractical. Harris likes to portray himself as a purveyor of Cold Hard Truths about matters both metaphysical and historical, but when it comes to the crimes of the American Empire, he can never quite bring himself to believe that things can be as bad as critics claim. In fact, his certainty that the worst that can be said about American foreign policy is that "idealistic" intentions were poorly executed is a fundamental axiom of his belief system. Ironically, then, Harris, who presents himself as being above history, is, like all of us, completely a product of the context in which he was born, lives, and works—the US Empire.

Harris' faith in Empire, his "philosopher's penchant" for imaginative speculation (which always seems to correlate with actually existing war crimes), and his almost childlike lack of information came brutally to light in his discussion with noted academic and dissident Noam Chomsky. The conversation between Harris and Chomsky played out over private email. (For another classic of the Sam Harris Makes An Ass Of Himself Over Email And Publishes The Results For Some Reason subgenre, see his exchange with Ezra Klein.)

The discussion begins with Harris expressing interest in a public debate. Chomsky responds with relatively polite disinterest (although he doesn't bother to disguise his distaste for Harris' antics):

Perhaps I have some misconceptions about you. Most of what I've read of yours is material that has been sent to me about my alleged views, which is completely false. I don't see any point in a public debate about misreadings. If there are things you'd like to explore privately, fine. But with sources.

In other words, *look, if you want to ask me a few questions about my writings I guess we can do that, but I don't have any particular interest in your half-assed critique of my life's work.*

Somehow, though, Harris seems to have taken this four-and-a-half-sentence-long email to be an open-ended invitation to have a long debate about morality, international relations, and philosophy. He starts his response email by encouraging Chomsky "to approach this exchange as if we were planning to publish it." (Word.) He appends several pages from *The End of Faith* in which he criticizes Chomsky for suggesting that the US bombing of the Al-Shifa pharmaceutical plant in the Sudan was the sort of crime that could be morally compared to the 9/11 terrorist attacks. Harris takes this to be an instance of "leftist unreason." Much of his discussion is devoted to Chomsky's belief that we should judge the actions of countries (and for that matter stateless terrorist networks) not by their stated intentions but by the predictable and predicted consequences of their actions. That Harris construes Chomsky's underlying ethics in an almost comically uncharitable way, inferring for some reason that Chomsky doesn't think that intentions can *ever* be morally relevant to anything, may ultimately be less interesting than the fact that he never gets around to answering the question with which Chomsky started his original discussion:

What would the reaction have been if the bin Laden network had blown up half the pharmaceutical supplies in the US and the facilities for replenishing them? We can imagine, though the comparison is unfair, the consequences are vastly more severe in Sudan. That aside, if the US or Israel or England were to be the target of such an atrocity, what would the reaction be? In this case we say, "Oh, well, too bad, minor mistake..."

When Chomsky pointed this out to Harris, the latter's excuse was that he hadn't read the discussion about Al-Shifa in the original source—Chomsky's *Radical Priorities*.

I have not read *Radical Priorities*. I treated your short book, *9/11*, as a self-contained statement on the topic. I do not think it was unethical or irresponsible of me to do so.

To put this in context, *9/11* was a 5 by 7-inch 128-page collection of essays and interviews in which Chomsky is interviewed *about* what he said in *Radical Priorities*. This is the equivalent of going to a major international debate about Marxism and only reading

The Communist Manifesto as preparation. (See Chapter 4.) It's that embarrassing.

When Chomsky eventually drags an answer out of Harris, things get pretty bad. Harris lays out a torturous scenario in which Al-Qaeda would have been justified in blowing up half of the pharamacuetical supplies in the United States:

> Imagine that al-Qaeda is filled, not with God-intoxicated sociopaths intent upon creating a global caliphate, but genuine humanitarians. Based on their research, they believe that a deadly batch of vaccine has made it into the US pharmaceutical supply. They have communicated their concerns to the FDA but were rebuffed. Acting rashly, with the intention of saving millions of lives, they unleash a computer virus, targeted to impede the release of this deadly vaccine. As it turns out, they are right about the vaccine but wrong about the consequences of their meddling—and they wind up destroying half the pharmaceuticals in the US.

Chomsky pours well-deserved scorn on this implied defense of the bombing of Al-Shifa. The same Clinton Administration that bombed Al-Shifa spent 8 years enforcing sanctions on Iraq that killed half a million children. (Notoriously, Clinton's Secretary of State Madeline Albright told CBS' Lesley Stahl that these deaths were "worth it.")

> And of course they knew that there would be major casualties. They are not imbeciles, but rather adopt a stance that is arguably even more immoral than purposeful killing, which at least recognizes the human status of the victims, not just killing ants while walking down the street, who cares?

In response, Harris does what Harris does—he retreats to the claim that realism is irrelevant because he was just engaging in a "thought experiment." As we've seen, he doesn't use the phrase "thought experiment" in a way that would be recognized by Judith Jarvis Thomson, Descartes, or any other philosopher who constructed interesting arguments with thought experiments at their center. Instead, his working definition of the phrase seems to be *some bullshit I wrote that should be immune from criticism.*

More tellingly, though, is the fact that Harris just can't bring himself to acknowledge that US intentions could be less than noble. If American bombs kill a lot of civilians, that must be because the country is acting for a good reason and, unfortunately, technology hasn't produced more precise military instruments. That killing a lot of people in order to cause fear and chaos could be part of US strategy—"shock and awe"—is unthinkable. He even finds a way to use the example of a fairly low-tech American atrocity—the My Lai massacre in Vietnam— as evidence of Americans' fundamental goodness.

[While this was] as bad as human beings are capable of behaving…what distinguishes us from many of our enemies is that this indiscriminate violence appalls us. The massacre at My Lai is remembered as a signature moment of shame for the American military. Even at the time, US soldiers were dumbstruck with horror by the behavior of their comrades. One helicopter pilot who arrived on the scene ordered his subordinates to use their machine guns against their own troops if they would not stop killing villagers.

Even putting aside that My Lai was hardly as rare an occurrence as the attention it got might lead you to believe—anyone with any doubt on this point should read the grisly transcripts of the Winter Soldier hearings organized by Vietnam Veterans Against the War—this gloss on the massacre is astonishing. *Current Affairs* editor Nathan J. Robinson lays out the historical distortions in his critique of Harris, "Being Mr Reasonable."

First, the helicopter pilot Harris mentions was Hugh Thompson, Jr., and far from representing the American moral mainstream, Thompson was ostracized and condemned by his fellow soldiers for his intervention in the massacre. In fact, popular opinion was overwhelmingly on the side of William Calley, the lieutenant who had ordered the killings. There were pro-Calley sympathy marches across the country, and the White House was flooded with calls for his release. A song called "The Battle Hymn of Lt. Calley," honoring the man who had ordered the execution of dozens of Vietnamese children, sold a million copies. Out of 26 soldiers initially charged with offenses related to the massacre, only Calley was convicted. But there was such a public outcry over the conviction that Richard Nixon reduced the sentence, and Calley ended up serving three years under house arrest, the only punishment handed out for a mass rape and the systematic murder of approximately 400 unarmed Vietnamese peasants.

In the original passage from *The End of Faith* in which he expresses outraged astonishment about Chomsky's failure to comprehend American goodness, Harris says that "not all cultures are at the same stage of moral development." He spends a fair amount of time in this passage—as is typical for Harris—congratulating himself on having the courage to formulate such a bold opinion. "It is time to admit," Harris sputters, that this "is objectively true," even though saying it is "radically impolitic." This is ridiculous for two main reasons. First, Americans have been making this argument since they began displacing indigenous peoples in the eighteenth century. Second, hard-core moral relativists who believe that it's never acceptable to criticize the injustices of another country or culture are a lot thinner on the ground in real life than they are in the writings of people like Harris, who are always congratulating themselves on their rejection of moral relativism. I imagine that if people at dinner parties react badly to Harris saying morality isn't relative, that's because they know the racist path on which he is treading. Keep in mind, again, that Harris wrote this during the Bush years, when the United States was waging multiple wars in the Middle East and justifying its actions, at least partially, on "humanitarian" grounds, claiming that the country was fighting wars of "liberation" to, for example, save Afghan women from the oppression they suffered in a morally undeveloped culture that forced them to wear burqas.

A clue about Harris' *actually* controversial views comes a couple of sentences later, when he suggests that we express such differences by saying that "not all societies have the same degree of *moral wealth*." This kind of wealth, he pontificates, tends to be a result of "political and economic stability, literacy" and "a modicum of economic equality." All of this sounds suspiciously like a description of the distinction between the Global North and the Global South—and a justification for the domination of the former over the latter. When you put this story about "moral wealth" together with the views he expressed in his exchange with Chomsky, the racist and imperialist story becomes fairly obvious.

From 1823 to 1857, the moral philosopher John Stuart Mill had a day job as an employee of the East India Company, helping to administer the subcontinent from an office in London. His 1858 book *On Liberty* was a classic defense of human freedom...or at least, freedom for *some humans*. Mill, unlike Harris, was at least open about who didn't quite count:

I hardly need say that this doctrine is meant to apply only to human beings when they have reached the age of maturity. We aren't speaking of children, or of young persons below the age that the law fixes as that of manhood or womanhood. Those who still need to be taken care of by others must be protected against their own actions as well as against external injury. For the same reason, we may leave out of consideration those backward states of society in which the race itself may be considered as not yet adult. The early difficulties in the way of spontaneous progress are so great that there is seldom any choice of means for overcoming them; and a ruler full of the spirit of improvement may legitimately use any means that will attain an end that perhaps can't be reached otherwise. Despotism is a legitimate form of government in dealing with barbarians.

Here's Sam Harris in 2014 discussing Israel's treatment of the Palestinians, making a similar argument about the necessity of brutality:

> Needless to say, in defending its territory as a Jewish state, the Israeli government and Israelis themselves have had to do terrible things.

Notice that Israel, like the United States, *has to* do these things; there is no moral question. In contrast, somehow Palestinians— or Iraqis or Iranians or Pakistanis or any Muslim-majority country—never *have to* do anything to Israelis or Americans. Instead, when Muslims "do terrible things," they do so of their own free will—even though Harris doesn't believe in free will— and as a pure result of "bad ideas" and lack of "moral wealth." He goes on:

> More civilians have been killed in Gaza in the last few weeks than militants. That's not a surprise because Gaza is one of the most densely populated places on Earth. Occupying it, fighting wars in it, is guaranteed to get women and children and other noncombatants killed. And there's probably little question over the course of fighting multiple wars that the Israelis have done things that amount to war crimes. They

have been brutalized by this process—that is, made brutal by it. But that is largely due to the character of their enemies... One of the most galling things for outside observers about the current war in Gaza is the disproportionate loss of life on the Palestinian side. This doesn't make a lot of moral sense. Israel built bomb shelters to protect its citizens. The Palestinians built tunnels through which they could carry out terror attacks and kidnap Israelis.

As usual, Harris ignores the real world and has no sense of—or interest in —the complexity of Palestinian politics and society. Like Mill writing about India, Harris paints all Palestinians with as broad a brush as possible and laments the "backwards state" of the nation as a whole. In other words, he totally flattens Palestinian reality. But, of course, Harris' complete ignorance doesn't stop him from making sweeping moral judgments, as when he says that Palestinians don't build shelters because they aren't morally developed enough to want to do so. (Never mind the Gazan schools and shelters that were built by the United Nations and destroyed by Israel, or the fact that Israel is far richer and has access to far more sophisticated weapons than Hamas.) In Harris' writings, Israel plays the role that Britain did in Mill: a good country forced to act brutally due to the moral immaturity of a barbaric people.

This is nothing more and nothing less than a mythologized, racist, and imperialist defense of an immoral hierarchy—in this case the domination of colonial powers over "morally impoverished" backwaters. The same penchant for defending hierarchy manifests itself in Harris' attempt to naturalize disparities between white and black Americans.

A pretty obvious explanation of those disparities in income, education, imprisonment, and the rest goes like this:

The class structure of American capitalism has been racialized from the beginning of our society. The ancestors of the overwhelming majority of black Americans were initially brought to America as slaves. Then, for the vast majority of the time between Emancipation and the present, most of them lived under a system of explicit, legally enforced racial apartheid. As late as the early 1960s, in much of the country, black people in much of the United States who tried to vote in elections or go to traditionally white universities to which they'd been admitted due to the intervention of courts were subject to mob violence with the complicity of state authorities. Since then, attempts to undo the effects of all of this horrific history have mostly been pretty lackluster. Even the idea of reparations, for example, has been relegated to the political margins. Meanwhile, in areas from housing to policing to economic austerity, right-wing and neoliberal policies have done a lot to re-enforce the disparities. As such, their continued existence is pretty unsurprising.

Seems pretty straightforward, right?

Well, an alternative explanation is given in Charles Murray and Richard Hernstein's 1994 book *The Bell Curve*. The larger thesis of the book is that economic inequality in general is largely attributable to genetically innate differences in IQ. In the most controversial chapter, Murray and Hernstein extend this thesis to economic inequality between the races. It says a lot about the dismal political environment of America in the 1990s that this was the part of the book in which Murray and Hernstein were perceived as having crossed the line. Their larger claims about poor and working-class people of all races didn't cause nearly as much commotion. While in what follows I'll talk about the debate as it's played out, which has focused on the racial dimension, it's worth underlining that *The Bell Curve* attempts to justify extreme inequality of all kinds as a result of natural forces. The book is in many respects the ultimate example of the intellectually limiting and morally damaging effects of naturalizing instead of historicizing.

Far from having some well-worked-out objection to the

obvious explanation of disparate racial outcomes, Murray has dismissed this explanation in remarkably vapid ways. In a debate with James Flynn, a figure I'll return to in a moment, Murray actually said that "most of the juice" had come out of environmental explanations "by the nineteen-seventies." Take a beat and let the absurdity of that sink in. Murray is saying that a single decade after the "Whites Only" signs came out of the restaurant windows and black people started to be allowed to vote, structural racism stopped explaining much of anything.

Sam Harris decided that the obvious explanation is wrong and Murray is right about all of this—that just as he himself is an unfair victim of mischaracterization because of his "philosopher's penchant" for exploring "corner cases," Murray is an unfairly maligned researcher who is simply following the empirical evidence wherever it may lead.

Interestingly enough, Harris hasn't always held this position. Or if he has, he hasn't always been comfortable expressing it.

In early 2019, *The Four Horsemen: The Conversation That Sparked an Atheist Revolution* hit the shelves of bookstores. This was long enough after the heyday of New Atheism that it made its way pretty quickly to the discount shelves. The book is padded out with introductions by all three of the remaining horsemen, but the bulk of it is a transcript of a "conversation over cocktails" (if you've ever watched my shows, you can imagine my tone of voice as I read out that description—my eye-roll as well) between Dennett, Dawkins, Hitchens, and Harris in 2007. The first three are at least occasionally witty or insightful, but most of the conversation is exactly the festival of undeserved self-congratulation that you might imagine. Still, it's an interesting time capsule. For example, there's a passage in which Dawkins says that it would be "good" to have "someone from the political right" in their ranks so people wouldn't think they had to agree with liberal politics to be an atheist. (In 2019, that's pretty clearly Not A Problem.) There's also a retroactively interesting moment when the four of them start discussing the idea that something could be true but too dangerous to put into print. They all agree that their critiques of religion don't fall into this category, but they discuss other possible scenarios in which the issue might arise.

DAWKINS: […] One should never do what some politically motivated critics often do, which is to say, 'This is so politically obnoxious that it cannot be true.'
DENNETT: Oh, yes.
DAWKINS: Which is a different—
DENNETT: Which is a different thing entirely. No, no.
HITCHENS: It would be like discovering that you thought the bell curve on black and white intelligence was a correct interpretation of IQ. You could say, 'Now what am I going to do?' Fortunately, these questions don't, in fact, present themselves that way.

Harris speaks next. Instead of challenging Hitchens on *this* point, he redirects the conversation to the comfortable terrain of Islam-bashing. However, in an infamous 2018 debate with Ezra Klein, Harris argues that the question about IQ does in fact present that way. He takes it for granted that the evidence is on Murray's side, and that everyone who disagrees is just too afraid of being called a racist to admit the truth.

In a 1996 *Vanity Fair* piece mocking MENSA, Hitchens notes that a propensity for crackpot reactionary and even outright fascist ideas "keeps popping up like a jack-in-the-box in the turgid writings of the IQ-obsessed." Even in the post-Iraq-invasion years when his liver and his moral core were both failing, Hitchens might well have had the sense to question the core premise that "intelligence" is measurable by IQ tests.

Stephen Jay Gould's brilliant 1981 book *The Mismeasure of Man* is essential reading on this point. Gould has a lot of fun with the history of attempts to "scientifically" validate hierarchy, like the efforts of nineteenth-century "race scientist" Samuel Morton. Gould discusses the discrepancies between skull-measurements Morton produced in 1839 and 1849, and concludes that Morton was so desperate to believe in white superiority that he unconsciously manipulated his data.

In 2011, Nicholas Wade published an article in the *New York Times* entitled "Scientists Measure the Accuracy of a Racism Claim." In it, Wade reports a new study from the University of Pennsylvania that purported to show that Gould's analysis of Morton was itself an example of unconscious bias corrupting data. A lot of reasonably well-informed people probably think that was the end of it. The *New York Times* never returned to the subject of Wade's article. (Wade would go on to publish an extremely sketchy 2014 book *A Troubling Inheritance: Genes, Race and Human History*, which was trashed even by some of the scientists whose research Wade cited in the book.) And as recently as March 2019, the always—"race science"—friendly *Quillette* published an article by Russell T Warne that recapped the Morton/Gould controversy without addressing the further developments Matthew Lau had summarized 2 years earlier in his excellent *Jacobin* article, "Remeasuring Stephen Jay Gould."

[M]ore recent evidence suggests that the reanalysis of Morton's skulls makes computational mistakes that favor Caucasians. And as several studies now show, the scientists did not ultimately challenge Gould's main claim that the inconsistencies between Morton's measurements in 1839 and 1849 indicate unconscious racial bias. Moreover, the differences between mean values for all races when corrected were, as Gould originally argued, so small as to be statistically insignificant.

This is all worth underlining, not because the Harrises and

Murrays and Wades of the contemporary world literally believe that intelligence can be measured the way that Morton thought it could, but because it's a revealing instance of a historical trend. Attempts to "scientifically" justify racial hierarchy always end up in the historical dumpster of empirically refuted garbage science.

Even when we take race out of the equation, the idea that intelligence is precisely measurable—either with calipers or with IQ tests—is dubious at best. "Intelligence" is a loose and poorly defined collection of abilities, ranging from mathematical acumen to business savvy, that don't always go together. Anything as simple as an IQ test must arbitrarily select some of these skills. I agree with the contrarian intellectual Nassem Taleb, who has argued that IQ is basically a self-licking ice cream cone for the testing industry. At best, it measures a certain kind of bureaucratic competence or demonstrates the kind of thing that would show up in a clearer way in diagnoses of severe mental impairments. As a Marxist, my take is that such tests have historically been geared toward job skills relevant in a certain stage of the development of industrial capitalism. No doubt as industry-driven proprietary testing's methodologies evolve it will come to reflect skill sets more closely attuned with the demands of the Silicon Valley economy.

Even when it comes to these narrow and arbitrarily selected subsets of "intelligence," we shouldn't accept the skull-measurer's conclusions. Some useful pushback to the Murray/Hernstein/Harris picture has been provided by the geneticist Eric Turkheimer, who I've had the pleasure of interviewing. His research strongly suggests that the causal connection between poverty and IQ test scores runs in the other direction. Siddhartha Mukherjee summarizes this research in *The Gene: An Intimate History*:

[I]n the 1990s the psychologist Eric Turkheimer strongly

validated this theory by demonstrating that genes play a rather minor role in determining IQ in severely impoverished circumstances. If you superpose poverty, hunger, and illness on a child, then these variables dominate the influence on IQ. Genes that control IQ only become significant if you remove these limitations.

This research is an elegant counterweight to the rebranded social Darwinism of *The Bell Curve*. Harris, of course, feels comfortable dismissing it, despite the fact that Turkheimer is a subject matter expert and Harris is not.

A key moment in the Harris/Klein debate is when he tells Klein that if environmental factors were driving things more than genetics, we would expect IQ differences to disappear when we look at the scores of *middle-class* black and white test-takers. Klein makes the obvious empirical point that middle-class African-Americans are much more likely to live in areas still affected by structural poverty, than their white counterparts.

The geneticist James Flynn also undermined biological determinism by demonstrating that IQ across populations has gone up as societies have grown more complex. Possible explanations of the effect have to do with education, nutrition, and the presence of more stimulating environments. This is called "the Flynn Effect" and it led to a priceless moment in the otherwise laborious Harris/Klein debate. Harris, no doubt relying on his meditative psychic powers, told Klein that he was not accurately representing Flynn's views. Klein responded that he'd just spoken to Flynn the day before. Harris, without missing a beat, continued to insist that Klein (and by extension Flynn..?) was misrepresenting Flynn.

Klein repeatedly tried to point out that Murray falls into a long history of white commentators trying to find "scientific" cover for their racial attitudes. While he doesn't entirely spell out this argument, Klein's point seems reasonably clear: This is a conclusion that beneficiaries and defenders of such hierarchies are clearly motivated to convince themselves is true. Past versions of this idea look like obvious nonsense from our historical vantage point. All of that gives us at least some inductive reason to suspect that Murray and Hernstein will turn out to be more of the same—especially given the painfully obvious alternative explanation of the data spelled out above.

Rather than engaging with this argument, Harris self-righteously sputters that "the history is completely irrelevant." As we'll see when we look at Shapiro and Peterson in the coming chapters, this might as well be the official slogan of the IDW.

Chapter Three

The Dragon Who Didn't Do His Homework

Jordan Peterson has a dark vision. Freedom itself—the Inheritance of the Western World—is under threat from "neo-Marxism" and the tyranny of collectivism. Hordes of college students demand gender recognition and express concern over inequality and the environmental crisis when they should be disciplining themselves (or at least tweeting in defense of their intellectual guru). The preoccupations with gender, savvy use of social media, and the advice to stand up straight and clean your room launched Jordan Peterson into the stratosphere in 2017 and 2018.

This is not to say that *literally* cleaning your room is a bad idea, and ten of his 12 rules amount to good advice.

Rule 1:

Stand up straight with your shoulders back

Rule 2:

Treat yourself like someone you are responsible for helping

Rule 3

Make friends with people who want the best for you

Rule 4:

Compare yourself to who you were yesterday, not to who someone else is today

Rule 7:

Pursue what is meaningful (not what is expedient)

Rule 8:

Tell the truth – or, at least, don't lie

Rule 9:

Assume that the person you are listening to might know something you don't

Rule 10:

Be precise in your speech

Rule 11:

Do not bother children when they are skateboarding

Rule 12:

Pet a cat when you see one in your street

The exceptions are Rule 5, "Do not let your children do anything that makes you dislike them," which has an authoritarian ring to it, and most especially Rule 6, which is a strong example of Peterson allowing his right-leaning politics to bleed into his life advice, and which simply states: "Set your house in perfect order before you criticize the world."

The problem, as Slavoj Zizek pointed out to Peterson in their debate, is that sometimes people's houses aren't in order precisely *because* of the condition of the world. As we'll see, this is the central contradiction in Peterson's worldview. On the one hand, he's a cultural conservative bemoaning disorder and social atomization. On the other hand, he's an uncritical apologist for the "free market" system that generates the very conditions he laments as predictably as a boiling tea kettle generates steam.

Peterson himself could do a better job of following Rule 10: "Be precise in your speech."

Here he is being imprecise in *12 Rules for Life*:

There are now whole disciplines in universities forthrightly hostile to men. These are areas of study, dominated by the postmodern/neo-Marxist claim that Western culture, in particular, is an oppressive structure, created by white men to dominate and exclude women (and other select groups); successful only because of that domination and exclusion.

Any reader who knows anything about any of these subjects should wonder what any of this has to do with either Marxism ("neo-" or otherwise) *or* postmodernism. Marxism doesn't explain gender oppression or anything else in terms of bad people designing structures with the conscious intention of screwing over people who are unlike them. Instead, Marx and Engels offer an analysis of such structures that's rooted in an economic and social analysis of the social relations that constitute and produce our material conditions. Postmodernism, meanwhile, is a school of thought that developed in rebellion against "grand narratives" about history and society—grand narratives like Marxism—and which argues there's no such thing as a stable truth.

While many people have tried to point this out to Peterson in the last few years, he continues to violate both Rule 10 and Rule 9 ("Assume that the person you are listening to might know something you don't.") as he conflates Marxism and postmodernism in a wildly imprecise way. In the years since he wrote that passage, Peterson has become even more slippery in his analysis. He often drops the "/" and sometimes even the "neo-" in "postmodern/neo-Marxist." He incoherently talks of "postmodern Marxism." In a video entitled "The Fatal Flaw in American Leftist Politics," Peterson goes so far as to claim that "French intellectuals" like Michel Foucault and Jacques Derrida, who had been Marxists and indeed "student revolutionaries in 1968," pulled off a "sleight of hand and transformed Marxism into postmodern identity politics" in the late-1960s when revelations about how bad the Soviet Union was made it "impossible to be a thinking and conscious person and to be Marxist."

This is embarrassing and historically illiterate. The most important revelations about the ugly realities of the Soviet Union came in Khrushchev's speech to the 20th Party Congress in 1956, when the Soviet leader lambasted both Stalin and Stalinism. Khrushchev's denunciations, combined with the Soviet invasion of Hungary later that year, led to mass defections from communist parties all around the world. No particularly new information about the Soviet Union or

China came out in the late 1960s, though, and when what had happened in Hungary in 1956 repeated itself in Czechoslovakia in 1968—with Soviet tanks rolling into Prague to crush another experiment in building a more humane and democratic form of socialism—it led communist parties in Western Europe to distance themselves from the Soviet Union and drift toward what would become "Eurocommunism." Meanwhile, a number of prominent Marxist intellectuals, including Jean-Paul Sartre, had long maintained their independence from the Communist Party. In fact, Sartre wrote a book in 1956 denouncing the Soviet invasion of Hungary, though this didn't mean he stopped being a Marxist.

The idea that there were Marxist intellectuals and movements that consciously separated themselves from the Soviet Union, its brutalities, and its associated parties doesn't seem to have occurred to Peterson as a possibility. This might be why he didn't notice that the Communist Party of France (PCF) was actually opposed to the uprising of French workers and students in 1968. Marxist student revolutionaries who participated anyway were by definition independent of the PCF and its pro-Soviet party line.

Meanwhile, Peterson gets a number of his facts wrong. To take one example, Foucault, far from being a protesting student in 1968, was by then a well-established intellectual in his forties. His first major book, *Madness and Civilization*—an important text in the development of postmodernism—was published in 1960. Similarly, Derrida's *Of Grammatology* came out in 1967—when the scholar was in his late thirties.

Peterson's cartoonishly conspiratorial story about the relationship between Marxism and postmodernism could hardly be less accurate. In 1966, Sartre called Foucault, "the last barricade the bourgeoise can still erect against Marx." Foucault returned fire, joking, "Poor bourgeoise. If they need me as a barricade, they have already lost power!"

That's funny, but the more we look at Foucault's political development, the more it seems like Sartre may have had a point. In an interview with the French journal *Ballast*, translated into English for *Jacobin* by Seth Ackerman, Daniel Zamora talks about how Foucault "was highly attracted to economic liberalism" in his later life, seeing in it a "much less bureaucratic" and "much less disciplinarian" form of politics than that offered by "the socialist and communist left, which he saw as obsolete."

There's also the question of what any of this has to do with "identity politics." Postmodernism is all about focusing on the particular in order to deconstruct "totalizing narratives." Feminists seeking greater gender equality, or trans activists working for equal rights and dignity, necessarily focus on more general categories. And unlike postmodernists, they have specifically political aims.

Nevertheless, Peterson's obsession with what he sees as the totalitarian Marxist aims of "postmodern identity politics" has defined his public career. Hardly anyone read his first book, *Maps of Meaning*—a dense tome that often reads like a psychedelic collaboration between Carl Jung and L. Ron Hubbard. *12 Rules for Life* stood out on shelves already full of similar self-help advice books because Peterson had become famous for his viral YouTube videos that warn of the threat "political correctness" poses to "Western Civilization." Specifically, it was Peterson's concerns about Bill C-16, *"An Act to Amend the Canadian Human Rights Act and the Criminal Code,"* which made him internet famous.

Peterson claimed that the bill was a "compelled" speech law under which he could be legally penalized for failing to use trans students' preferred pronouns. Here are several examples of criticisms he made during his crusade:

I've studied authoritarianism for a very long time—for 40 years—and they're started by people's attempts to control the ideological and linguistic territory...There's no way I'm going to use words made up by people who are doing that— not a chance.

There's only two alternatives...One is silent slavery with all the repression and resentment that that will generate, and the other is outright conflict. Free speech is not just another value. It's the foundation of Western civilization.

These laws are the first laws that I've seen that require people under the threat of legal punishment to employ certain words, to speak a certain way, instead of merely limiting what they're allowed to say.

I can envision a student or a colleague insisting that I call them using gender neutral pronouns. "Zhe" or "Zir." ...I think that those gender neutral pronouns are...connected to an entire underground apparatus of political motivations...I think that uttering those words makes me a tool of those motivations...the mouthpiece of a murderous ideology.

In liberal democratic societies there are often competing interests and even conflicting rights.

Since Bill C-16 has passed no one in Canada has been fined for using the wrong pronoun. Just how a potential conflict between the human rights code and Section 2 of the Canadian constitution (the section establishing the right to free expression) will be worked out in the courts remains to be seen. What is immediately obvious, however, is that Peterson's claim that Bill C-16 amounts to an attempt to use an "underground apparatus," an apparatus apparently built by a group of malicious neo-Marxists, to enforce a chaotic and resentful ideology and destroy the West, is not just imprecise, but unhinged. Bill C-16 is, like any other attempt to establish and protect human rights, a product of the liberal and democratic principles that Peterson would associate with the West.

Peterson's battle with what he imagines to be the

agenda of "the radical left" simply can't be taken seriously, especially when you consider his self-confessed ignorance. At the beginning of his debate with Zizek, Peterson made the shocking admission that, despite the fact that he has built his fame by decrying the dangers of something he calls Marxism, prior to his preparation for the debate he hadn't reread *The Communist Manifesto* since he was 18 years old.

If he hadn't felt moved to pick up Marx's shortest and most popular book at any point in the last few decades—and to put this into perspective, there are editions of the *Manifesto* that are padded out with multiple prefaces, introductions, and explanatory essays that still manage to be thinner than the book you're reading now—it's safe to say Peterson hasn't been poring over the three volumes of *Capital*. Even when it comes to *The Communist Manifesto*, he seems to have missed Marx's (and Engels') point by a mile. In his debate with Zizek, Peterson claimed that "even Marx" "admits" in the *Manifesto* that capitalism is the most productive economic system that has ever existed. Benjamin Studebaker sets him straight on this point in an article in *Current Affairs*.

> This is the point in the talk where Peterson most clearly reveals his lack of engagement with the content of Marx's theory of history. Marx thinks that each economic system is the most productive in history when first introduced, but that eventually each outlives its usefulness and is replaced by something more appropriate to the technology of the time.

That Peterson has a shaky grasp on history and Marxist theory is to be expected. As I've been saying throughout this book, the Petersons of the world want to *naturalize* or *mythologize* the injustices we see around us instead of analyzing them as a function of historical processes that, because they are human-made, can be rectified in the future. Unsurprisingly given his adherence to Jungian psychology, Peterson's particular specialty is mythologizing. For example, instead of just saying that feminists or trans activists are a threat to traditional values, he'll go on and on about defending capital-O Order against "the dragon of chaos." Similarly, he likes to accuse those who desire an "equality of outcome" of having an "incomplete" view of the world because they see social systems through the mythic archetype of "the Tyrant" without understanding that the latter is "forever bound together" with the archetype of "the Wise King."

Peterson also does plenty of naturalizing—in fact, he's so fond of it that you can see him on YouTube declaring "I'm an evolutionary biologist" even though he holds no degrees in this field. One of his favorite arguments is that "the radical left" is foolish for trying to transcend hierarchy given that *even lobsters have hierarchies*. The argument goes like this: Given how long ago the ancestors of humans and lobsters diverged, a tendency to arrange ourselves into dominance hierarchies must be etched deep into our biology.

This is a completely vacuous assertion that could be used to defend *any* hierarchical system at any point in history. An antebellum Jordan Peterson could have used it to defend the hierarchy of slaves and plantation-owners (and in fact, many did affirm that slavery was "natural" and evident in such important texts as the Bible). An eighteenth-century Peterson could have used it to defend the hierarchy of King, Lords, and Commons. Meanwhile, no one really suggests that literally *all* forms of social interaction that could be described as "hierarchical" are problematic. I've been a socialist for a long time now, and I've never once heard anyone claim that parents and toddlers should have a completely equal say on matters such as bedtime and wandering into traffic—and for that matter I don't want each NBA team to take turns winning a championship. The question is whether any *particular* hierarchy is justified, and if not, whether and how we can get rid of it.

Putting aside Peterson's arguments about Hierarchy in General, we should ask whether the hierarchical structure of capitalist businesses can or should be changed. To answer *that* question, we must move away from bizarre Aesopian fables and take a look at the real world. The Mondragon Corporation is both a worker-owned cooperative and one of the most successful businesses in Spain. The Marxist economist Richard Wolff visited Mondragon in 2012, and here's what he reported:

> Given that MC has 85,000 members...its pay equity rules can and do contribute to a larger society with far greater income and wealth equality than is typical in societies that have chosen capitalist organizations of enterprises. Over 43% of MC members are women, whose equal powers with male members likewise influence gender relations in society different from capitalist enterprises.
>
> MC displays a commitment to job security I have rarely encountered in capitalist enterprises: it operates across, as well as within, particular cooperative enterprises. MC members created a system to move workers from enterprises needing fewer to those needing more workers—in a remarkably open, transparent, rule-governed way and with associated travel and other subsidies to minimize hardship. This security-focused system has transformed the lives of workers, their families, and communities, also in unique ways.

Wolff emphasizes that Mondragon is not a utopia. It's a work in progress with problems and limitations and internal contradictions to be overcome. It's hard to deny, though, that its considerable success demonstrates that the rigid economic hierarchies that emerge from the separation of labor and ownership can be transcended without this leading to economic collapse or famine or the dragon of chaos terrorizing the countryside.

I could keep going in this vein, but dealing with Peterson adequately requires more than just listing off the many, many things he gets wrong or highlighting his dangerous contradictions—for example, obsessing about "freedom" while getting chummy with authoritarians like Viktor Orban. An adequate left response to him must involve a much more focused and even empathetic look at the real cultural crisis to which he is speaking and which has led many sad young men to turn to him for comfort. In capitalism human beings are alienated, and the young and mostly (but not all) white men that Peterson speaks to are no exception.

The shifting demands of work, technology, and the economy have left many feeling confused and adrift, including some of those "succeeding" in the emerging paradigm. (At this point, one should ask whether working 90 hours a week to destroy the climate for Goldman Sachs is a "success" in any meaningful way.) Peterson is the perfect guru in the sense that his highly emotional, pop-culture-friendly, and middle-brow intellectualism, which is grounded in appeals to an imagined past, provides a perfect object of focus for this frustration. His style of communication is purely of the now, and the conservative currents he embodies have the perfect bedtime story appeal; they provide a clear, yet individualistic and capitalistic, path forward for the alienated young man.

One of the most dangerous things the left can do is to write off the demographic to which Peterson appeals because of its relative racial and gender privilege. For one thing, setting all else aside, young and angry white men have historically been a pretty dangerous group. They are the subset of the population most likely to become school shooters or join fascist movements. Far better that someone addresses their alienation in a constructive way and channels their justified frustration in a positive direction.

Analyzing Peterson's shaky assertions, sloppy and eccentric thinking, and disturbing generalizations is important. However, this critique clearly does address the core emotional and psychological needs that he is speaking to as a public figure.

Like everyone else, young white men are trying to muddle through life in a relentlessly complicated, unequal, and—to use one of Peterson's favorite terms—chaotic era. All the IDW and IDW-adjacent figures mentioned in this book are spreading narratives that soothe general anxieties by smoothing over the complications of real life.

It's easy to make lobster, crying, and cider jokes, and I'm certainly not above that. See the Cider Story with Joe Rogan video. Seriously, watch it. Oh my God that was amazing. But we need to develop a deeper response than that. The cosmopolitan socialist synthesis that I'm arguing for aims to deal head-on with the anxieties, pain, and confusion that Peterson evokes.

To begin framing the left's response, one must appreciate the degree to which Peterson is taking on the biggest macro-issues of our time and trying to solve them with the smallest self-help micro-solutions. Socialists can do better. We can start by analyzing the material roots of the uptick in alienation and despair that fuels Peterson's book sales. That doesn't mean telling people *not* to care about spiritual fulfillment or personal meaning. It's not an either/or. What we should point out, though, is that increasing numbers of younger people in our late-stage capitalist economy are pushed into forms of precarious freelance pseudo-entrepreneurship. People aren't having trouble maintaining relationships or waiting longer to have children because of Marxism or feminism or the existence of trans people; they are having trouble maintaining relationships or waiting longer to have children because they live lives defined by relentless anxiety and undercompensation. The loss of stability experienced by so-called "millennials" and "zoomers,' moreover, hasn't even been compensated by a meaningful increase in autonomy. Whether you're employed by General Motors or Uber, workplaces at the end of the second decade of the twenty-first century remain sites of autocracy.

The example of Mondragon suggests that there's a better way. Peterson claims to cherish the Western tradition, but he has no interest in applying the best of that tradition—the Enlightenment concepts of democracy and self-determination—to the workplace. Instead of asking people to simply clean their rooms while everything burns down around them, socialists must reimagine the oppressive and alienating systems that structure both the boardroom and the shop floor.

What about the issues of meaning and purpose that Peterson constantly evokes? I have no doubt that his willingness to speak so frankly to such fundamental parts of our lives is inseparable from his broader appeal. Again, though Peterson's approach is lacking, the themes he's dealing with should not be dismissed but should be addressed in more integrated and sound ways. The work of anthropologist Scott Atran and depth psychologist James Hillman provide potent contrasts to Peterson's mythology-enhanced market fundamentalism.

Unlike Peterson's dreck, Atran's ethnographic research on terrorism has actually been used in real-world situations, and not just in thought experiments and podcasts. Namely, what he's done is connect the rise and discourses of both jihadism and the alt-right to the alienation and lack of shared purpose endemic to hyper-consumerist societies like the United States of America and Saudi Arabia.

Consideration of this point leads us right back to the core contradiction in Peterson. Though he affirms that he's deeply disturbed by the alienation, fragmentation, and disruption of cultural continuity brought about by late capitalism, he nevertheless expresses an enthusiastic commitment to "free markets." This leaves him without any mode of response more useful than an addled combination of conspiracy theories and self-help advice about lobsters and chaotic women.

While I don't agree with Atran about everything, he approaches the problem of male alienation from the correct starting point, recognizing that fostering "sacred meaning" and "group bonds" in liberal market societies is a defining challenge of our era. Indeed, this is a challenge that international socialists should take up as they work to transform society.

If you're interested in contemporary Jungian psychology, you can do *much* better than Peterson. James Hillman is an archetypal psychologist who actually ran the Jung Institute in Vienna at one point in his career. He went on to co-author a brilliant book in the 1990s entitled *We've Had a Hundred Years of Psychotherapy and the World is Getting Worse* that offered a number of incisive points about modern day alienation.

We've had a hundred years of analysis, and people are getting more and more sensitive, and the world is getting worse and worse. Maybe it's time to look at that. We still locate the psyche inside the skin. You go inside to locate the psyche, you examine your feelings and your dreams, they belong to you. Or it's interrrelations, interpsyche, between your psyche and mine. That's been extended a little bit into family systems and office groups — but the psyche, the soul, is still only within and between people. We're working on our relationships constantly, and our feelings and reflections, but look what's left out of that...What's left out is a deteriorating world.

I feel comfortable with Hillman's arguments about mythology, archetypes, dreams, and spirituality because his analysis of these subjects is connected to a seriously grounded materialist understanding of how social and group practices form and influence psychological health. In an interview entitled "You're Not Paranoid—The Boss Really is Out to Get You," Hillman used everything from air pollution to homelessness as potent examples of problems that should in fact cause us psychic anguish, but that psychology, with its focus on the self, will not solve. Though Hillman never denied the role individual development, psychological maturation, and spiritual growth play in making a person content, he understood that a pained response is often a natural response to a shockingly unequal and ecologically poisoned world. Hillman's wasn't a bedtime story, but a real challenge to reorient ourselves to face the dissociations and collective trauma we are constantly causing and with which we are constantly dealing. Both Hillman and my own mentor Suzin Green (a strategist whose work and analytic insight has helped me greatly) also deal in ambiguity and in discerning underlying dynamics and challenges instead of rigid and linear readings and interpretations of mythological texts.

Peterson's artificial division between market mythology and reactionary conservativism will ultimately doom his project intellectually. Tragically, as a self-help source for confused and alienated people, he will likely continue to find a ready market for his wares unless we provide something more compelling. Wolff, Atran, and Hillman offer us a useful start, but it will be up to us to do the work.

Chapter Four

Ben Shapiro and Other Dishonorable Mentions

While Harris and Peterson position themselves as above-the-fray gurus who just happen to spend all their time defending existing hierarchies and attacking "the left," Ben Shapiro doesn't hide his right-wing ideological commitments.

Shapiro made his bones writing columns in support of the invasion of Iraq and before he joined the IDW he was known as a right-wing Wunderkind. Matt McManus aptly summarizes this early career in a review of Shapiro's new book *The Right Side of History: How Reason and Moral Purpose Made the West Great.*

The subjects of his early books, with cute titles like *Bullies: How the Left's Culture of Fear and Intimidation Silences Americans* and *How to Debate Leftists and Destroy Them,* seemed closer to Ann Coulter or pre-Trump Glenn Beck than Russell Kirk.

But, Shapiro's aim in the *The Right Side of History* is, at least in part, to rebrand himself in the image of his IDW colleagues. Rather than another disposable pundit who screeches conservative AM radio talking points (peppered in his case with superficial references to "logic" and "reason"), he now wants to be seen as someone who talks about big ideas like Religion and History and Greek Philosophy and the glories of "The West." When he points to the rise of political anger to illustrate the decline of Western culture, he even makes a few half-hearted stabs at telling a story that blames Both Sides. Needless to say, his story revolves around the sort of false equivalence between woke college students and alt-right neofascists you'd expect from the likes of Harris or Peterson.

In between an opening chapter on the relationship between happiness and moral purpose and a final chapter remonstrating the left for "returning to paganism," Shapiro spends most of his time on the greatest-hits of Western civilization: Aristotle, the Bible, the Founding Fathers, and the Enlightenment. All of this might seem to contradict my argument that what unites the IDW is their desire to ignore the historical origins of contingent social structures and instead naturalize and/or mythologize them. After all, by examining the history of the West isn't Shapiro historicizing the present?

Maybe, but let's take a closer look:

Shapiro claims that something called "Western civilization" owes its greatness to two sources— "Judeo-Christian" religiosity on the one hand and "Greek reason" on the other. Borrowing terminology from the German-Jewish exile philosopher Leo Strauss (who himself took it from the early church father Tertullian), Shapiro short-hands these influences as "Athens" and "Jerusalem." In Shapiro's telling, the apex of both influences was the founding of the American Republic.

> The philosophy of the founders, made material in the creation of the United States and in the continuing quest to fulfill their ideals, has been the greatest blessing for mankind in human history. The United States has freed billions of people; it has enriched billions of people; it has opened minds and hearts.

In short, Shapiro confuses a fable about America for real historical events and figures. He uses history as a prop as he reenacts his preferred mythology.

The notion of Western civilization has its own history, of course. In the United States the idea of Western civilization, specifically the idea to teach Western Civ as an introductory course in university, took hold between the two world wars. In an essay entitled "The Rise and Fall of the Western Civilization course" and published in *The American Historical Review* way back in 1987, Gilbert Allardyce described the invention of Western civilization as one of the great success stories in the history of the historical profession in America.

The idea had a good run, but as Allardyce also noted, by the end of the 1960s the notion had been overturned and scholars had come to see the Western Civ project as having been conditioned by two wars.

"Americans envisioned themselves as partners with European democracies in a great Atlantic civilization, formed from a common history, challenged by a common enemy, and destined to a common future," Allardyce writes.

Unfortunately, like any attempt to tell a sweeping story about history and congeal the past into a coherent narrative with fixed identities, the effort involved quite a bit of flattening of people, cultures, and ideas. For example, an ancient Jew or ancient Greek would have held more in common with an ancient Egyptian than they would with an ancient Celt, and certainly would have had more of a connection with the Egyptians than with twentieth-century Americans.

Simply put, Shapiro's idea of the "West" is an invention, a twentieth-century construction, and Shapiro's claims about "Judeo-Christianity" are, if anything, even more risible. Again, historians have shown that the idea of "Judeo-Christianity" didn't emerge until the mid-twentieth century at the earliest and for largely the same reasons that the Western Civ course arose. The more substantial concept, the construct scholars of religion call the Abrahamic religions—Judaism, Christianity, *and Islam*—unfortunately includes Muslims, the very category Shapiro is most concerned with excluding from "the West." Indeed, Shapiro's writing on Palestinians, Arabs, and the Muslim world in general has been so appalling that the easiest way to find a lot of this material now is to click through a post on his blog entitled, "So Here's a Giant List of All the Dumb Stuff I've Ever

Done (Don't Worry, I'll Keep Updating It)." Though someone who didn't know much about Shapiro might find that title charming, the things cataloged in it are anything but. Here's a sample:

1. A column he wrote advocating the "transfer" (read: ethnic cleansing) of all 4.5–5 million Palestinians from the West Bank and Gaza, plus the other million-and-a-half or so who live within Israel's pre-1967 borders and have Israeli citizenship.

2. In 2013, a "high-ranking" allegedly told Shapiro that Chuck Hagel had taken money to speak to a group called "Friends of Hamas." Apparently, Shapiro's paranoid Islamophobia had reached a point where that *sounded like something that might be true*, and so he printed it. Oops. Sorry, Chuck Hagel, the former Republican Senator from Nebraska who was an occasional voice of restraint against interventionism in the Bush era (although he did initially vote to invade Iraq in 2003) and who was nominated by Obama to be Secretary of Defense. That Chuck Hagel was indeed not an associate of Hamas. Shocker.

3. A column entitled "Enemy 'Civilian Casualties' OK By Me." Enough said.

4. In 2017, Republican Congressman Steve King tweeted about the birth rates of Muslim immigrants to Europe. Praising Dutch fascist Geert Wilders, King tweeted, "Wilders understands that culture and demographics are our destiny. We can't restore our civilization with somebody else's babies." Even though it's a pretty good bet that someone with King's worldview would also put Shapiro's family in the "somebody else" category, at the time Shapiro contorted himself into a rhetorical pretzel trying to find a non-racist way to interpret King's comments. Maybe, Shapiro said, King was just bemoaning the fact that these babies wouldn't be properly "assimilated" into "the West"

as they grew up. In Shapiro's mind, *that* would be a totally reasonable and non-bigoted thing to say. Two years later, Shapiro appended an update to the column, stating, "His later open embrace of the terms "white nationalist" and "white supremacist" suggest that the first interpretation described below was not as implausible as it seemed at the time." Word.

The point of highlighting these columns isn't to say that Shapiro can't learn or grow, but to underline the rhetorical game that he's playing. Shapiro wants to rebrand himself as a refined intellect — or, in the immortal and absolutely embarrassing words of the Failing New York Times, the "Cool Kids' Philosopher," but has he substantively changed any of his views on Palestinian human rights? US foreign policy? Pluralistic democracy? No.

Even if Shapiro no longer publishes pieces with titles like "Enemy 'Civilian Casualties' OK By Me," he still cheers on every new American or Israeli attack on an "enemy" nation. When it comes to Israel/Palestine, he still opposes *both* a one-state multiethnic democracy in which everyone has equal rights *and* a two-state settlement in which the Palestinian territories would become independent. Presumably, the residents of the West Bank and Gaza will never accept being a permanently stateless non-citizen underclass. So, what does Shapiro think should be done with them? These days, he just doesn't say, though we can infer what his feelings are given that he continues to defend this 2012 tweet:

Israelis like to build. Arabs like to bomb crap and live in open sewage. This is not a difficult issue. #settlementsrock

Indeed, in the same blog post where Shapiro lists off "dumb" things he now disowns, he has a separate list of "stuff the left is taking out of context." This tweet is the first item on that second list. He defends it by pointing to follow-up tweets in which he "clarifies" that he's only talking about "Arabs who participate in the Israeli/Arab conflict."

Even in the rebranding exercise that is *The Right Side of History,* the assumptions underlying Shapiro's ludicrous attempt to defend Steve King in 2013 are evident. A Muslim is "somebody else" who exists outside "our" civilization. Never mind that Judaism and Islam have at least as much in common with each other as either has with Christianity, or that the encounter between Islam and "Greek reason" actually happened a bit earlier than the parallel encounter between Christianity and the same Greek thinkers. In fact, if Muslim intellectuals hadn't preserved some of those texts, they wouldn't have been available to the likes of Thomas Aquinas.

Shapiro's approach throughout *The Right Side of History* is to randomly sprinkle in sentences referencing some of these historical realities—just enough so that no one can accuse him of ignoring them—while still centering his narrative on arguments that implicitly ignore these complications. He never explains why the magical combination of "Judeo-Christian" religion and "Greek reason" that makes "the West" great doesn't also make "Islamic civilization" great, or indeed why he believes these two "civilizations" (itself a very problematic term) are in conflict. Furthermore, in all his paeans to Judeo-Christianity, he never discusses that for much of the history of Jewish-Christian relations, the Jews were viciously persecuted. To take one particularly absurd example, Shapiro praises Immanuel Kant for developing the "closest thing" secular philosophy has ever come to "a serious sense of reason and purpose" without mentioning that Kant referred to Jews as "a nation of cheaters."

My point here isn't that Kant is canceled but that Shapiro's failure to wrestle with these obvious points betrays an intellectual shallowness—or perfidy—that completely undermines his ahistorical and rosy view of "Judeo-Christian"

"Western civilization." He is ultimately not a serious thinker, as his annoying ninth grade World History class way of Quoting Great Quotations should be enough to show (he always gives each figure's first and last name, and birth and death years, in a way that suggests he's laboriously copying this information from an encyclopedia). He's just not interested in history, or in a history that centers not on dates, but on the interpretation and analysis of highly complex causal processes. Instead, what Shapiro is interested in is in reinforcing his reactionary mythology while repackaging himself as a Deep Thinker.

However, Shapiro's attempt at rebranding went up in flames in May 2019, when he was interviewed by the BBC's Andrew Neil. A few minutes in, Neil notes that newly passed abortion laws in Georgia, "which you are much in favor of," would "seem to take us back to the dark ages." Under these laws, Neil noted, a Georgian woman who suffered a miscarriage could be sentenced to 30 years in prison. How does Shapiro's support for such "extreme hard policies" fit with his new, more reasonable, image?

Shapiro's response was to repeatedly stutter the word "sir," claim that his support for the law followed from "science" (because a fetus is a human life), and to call Neil a biased "liberal" for attacking "the pro-life position." This last part rings particularly hollow, since for decades members of the "pro-life" movement have insisted that only the doctors who performed abortions should be punished. When Donald Trump said in 2016 that there should be "some punishment" for the women who have abortions, even many pro-lifers considered this a gaffe. The needle is constantly moving, though, and Shapiro is far too loyal a conservative foot-soldier not to move with it.

Meanwhile, Shapiro's arguments about "science" are typical of his hacky and superficial attempts at providing "logical" arguments to alienated and sad teenagers. Ben Burgis devotes a chapter to Shapiro's confusion about this subject in his book

Give Them an Argument: Logic for the Left. Suffice to say that "science" can't answer normative questions, such as whether a woman's right to bodily autonomy outweighs a fetus' right to life, or indeed whether a literally mindless first trimester fetus is the sort of entity capable of having rights in the first place. These are moral questions, and Shapiro's strategy is to evade their complexities by talking quickly and loudly and projecting confidence. He's a two-bit pedant who, when cornered, has nothing to offer but bluster and self-victimization.

Neil started laughing when Shapiro asked him why he pretended to be an objective journalist rather than admitting to being a biased "liberal." Shapiro huffily insisted that it was a serious question. Neil responded, "If you knew how ridiculous that was, Mr Shapiro, you wouldn't have asked it."

Indeed. Neil is a Tory who backed the Iraq War, denies the scientific consensus on climate change, and has worked for *The Economist*, *The Sunday Times*, and the Conservative Party.

Neil then tried to explore Shapiro's history of making absurd and inflammatory statements, emphasizing the obvious point that Shapiro has done quite a bit over the years to create the kind of political culture he criticizes in the first chapter of the book. After a few questions along these lines, Shapiro stormed out of the interview.

As funny as that was to watch, it's important to draw out and emphasize a distinction I mentioned earlier. Even when it comes to characters as contemptible as Ben Shapiro, I don't think people should be condemned for positions from which they've truly moved on. Absolutely unforgivable pasts are few and far between, and our default reaction to people rejecting bad views should be to encourage their growth. If Neil *had* been shaming Shapiro for being wrong about things he was now right about... well, I might still find the clip funny, but I would agree that it was unfair. The problem is that Shapiro's big list of Stuff That No One Is Allowed to Criticize Me for Now Because I Called Take-

Backsies is almost a parody of genuine moral growth. It reminds me of nothing so much as the scene in Oliver Stone's *Nixon* in which Anthony Hopkins, playing the disgraced president, is reading transcripts of the White House tapes crossing things out and muttering to himself, "Nixon can't say that...it makes me sound anti-Semitic!"

If Shapiro ever experienced a real epiphany and decided to actually rethink the substance of his views and formulate new positions that, for example, took seriously the humanity of Palestinians and other Arabs, I'd still criticize him for the thousand other things he's wrong about, but I'd welcome him to the Basic Human Decency Club (at least when it came to Israel-Palestine relations). This is a point of principle and it's one worth spelling out explicitly.

The failure by some on the left to recognize that we should encourage moral growth instead of shaming and canceling people for having gotten things wrong in the past is a big part of the phenomenon Mark Fisher brilliantly described in his essay "Exiting the Vampire Castle." Fisher wrote the essay in 2013, 4 years before his suicide, but every time I read it, all I can think is that, apart from a dated example or two, it could have been written last week.

"Left-wing" Twitter can be a miserable, dispiriting zone. Earlier this year, there were some high-profile twitterstorms, in which particular left-identifying figures were "called out" and condemned. What these figures had said was sometimes objectionable; but nevertheless, the way in which they were personally vilified and hounded left a horrible residue: the stench of bad conscience and witch-hunting moralism.

Long before the right was dining out on endlessly clippable examples of collective performative browbeating, Fisher was outlining the serious political, emotional, and moral costs (and underlying neoliberal logic) of "call-out culture" — a culture that, unfortunately for all of us, is hardly confined to Twitter. What he called "the Vampires' Castle" is "driven by a *priest's desire* to excommunicate and condemn, an *academic-pedant's desire* to be the first to be seen to spot a mistake, and a *hipster's desire* to be one of the in-crowd."

What should be obvious is that all of these desires are best served by a left that remains, as it has for the last century of American history, a small and insulated subculture. Of course, those under the psychic sway of the Vampires' Castle do not consciously desire political marginality or believe they're engaging in moralistic self-policing, but this is exactly what is happening. And if it continues, the left will fail. As Fisher says, the Vampires' Castle "doesn't know how to make converts. But that, after all, is not the point."

I am not arguing that no one on the left has ever said or done anything racist or sexist or transphobic, or that we shouldn't care if they do. I'm also not claiming that we should disavow the historical importance of identity in favor of a simplistic economic reductionism that tells people not to worry about "merely cultural" issues. That's exactly the wrong way to fight the Vampires' Castle. What I am saying, however, is that the left (or, at least, the online left) suffers from a deficit of empathy, and we will continue to devour each other—and thus fail to win power in society—if we don't reject the confused moralism that permeates so much left-wing discourse.

Above and beyond everything else that's wrong with this kind of behavior, it's an ongoing gift to our political enemies. Claire Lehmann's magazine *Quillette* has printed nonsense about IQ that borders on a contemporary version of phrenology (the art and science of using skull-measurements to determine personality traits and intelligence). It's published ridiculous capitalist apologetics, including a crude article by an Amazon warehouse worker defending the company's disgusting labor practices. It's also published a few good articles by leftists like Ben Burgis and Matt McManus, who make a practice of going into enemy territory to argue for leftist ideas. Though I back them in their incursions, we should be crystal clear about *Quillette*'s reactionary editorial line.

But here's the thing. *Quillette* and similar magazines don't attract attention to their toxic material because there's a massive pre-existing audience for their worst takes. Rather, they generate their readership by publishing a never-ending stream of "oh my God, look at these leftists being crazy" articles. Does the right exaggerate and lie about these things? Sometimes. But these things do really happen from time to time as well, and when they do they cause real problems.

Entire careers are built on this nonsense. We wouldn't have ever heard of Bret Weinstein if not for what happened at Evergreen College. Far too much ink and podcast hours have been spent laboriously rereviewing his version of the Evergreen drama. Without attempting to relitigate the he-said/she-said of that situation this far out, I will say that I suspect that my friend and editor Doug Lain was correct to ask whether an increasingly corporatized administration was playing divide-and-conquer by deflecting student anger away from real decision-makers and onto a professor who was arrogant and tone-deaf enough about the students' legitimate anger and activism to make for an easy target. The way that activists turned their attention to this ridiculous and thoroughly unimpressive person (who none-the-less has connections) turned him into a *cause celebre* for the right. Whatever else is true about all this, what matters most is that Bret and his brother Eric Weinstein are here with us now, pushing the right-wing "classical liberal" pablum of the IDW.

In a world where the left was just a little bit better at acting strategically, Bret might well still be teaching biology at a nice hippie liberal arts college, while Eric quietly did whatever it was that he did for billionaire ghoul Peter Thiel. Instead, we have the Weinsteins to thank as much as anyone for the birth of the IDW brand. Bret's narrative about Evergreen is one of these guys' founding myths, and Eric is the one who came up with the absurd self-aggrandizing label "Intellectual Dark Web" in the first place. Bret's former colleague, Professor Nancy Koppelman, summarizes his post-Evergreen success in a Medium post entitled "Bret Weinstein's Second Act":

Before the events that led to his resignation from Evergreen his twitter page had few subscribers; now there are over 160,000. His Patreon account has over a thousand supporters who pay between $2 and $100 each month for exclusive access to him. He speaks to international audiences. He testifies before Congress.

That last detail is worth lingering on. Indeed, Weinstein has taken his aggrievement and actively supported Republican efforts to suppress free speech on college campuses. The IDW and the right in general love to have it both ways with free speech. On the one hand, if a reactionary is criticized for something they say, Free Speech is Under Attack. On the other hand, if a left-wing professor says something they find objectionable, or if too many faculty members have political views they dislike, they have no problem asking the government to step in to examine the curriculum and impose "balance." Heads, I win, tails, you lose.

That's about all that's worth saying about the Weinsteins in this "dishonorable mention," but Joe Rogan is a different beast. He could easily be hanging out with cooler people than Peterson, Shapiro, Harris, or the Weinsteins, and he sometimes does hang out with cooler people. While he's provided an uncritical platform for some people who truly suck—including several who I've written about in this book, he's recently sat down for interesting and in-depth interviews with Bernie

Sanders and Cornel West. His show has exposed a lot of people to left-wing ideas they wouldn't have heard anywhere else. It would be foolish to dismiss the appeal of Rogan's personality and platform, a mistake to ignore that he represents the most genuinely heterodox mash-up of politics and influences in the IDW.

And, finally, we come to Dave Rubin. About the only serious point to make about Dave—who as I'm writing this just announced that he was joining Glenn Beck's The Blaze—is that the IDW's endorsement of and cross-promotion with Rubin is a indication of the group's astonishing intellectual and moral bankruptcy. While Sam Harris will pal around with far-right figures like Charles Murray and has a worldview that is at bottom just as credulous and dim-witted as Rubin's, Dave is transparently and obviously stupid. I don't idealize or put on a pedestal those of us who talk for a living and I still find it genuinely astounding that Dave's mind-blowingly insipid, self-contradictory recitation of a handful of right-wing homilies and talking points has sustained a career.

Still, right-wing media is one of the easiest gigs in the world.

Chapter Five

Beyond the IDW

The answer to the IDW and the new right in general is an Internationalist-socialist synthesis that is all about global and materialist politics. The alienated and confused young men who flock to someone like Jordan Peterson obviously won't be won over to the left by telling them constantly to acknowledge and question their privilege. Nor can the even greater mass of people who aren't as plugged into politics or the culture war as IDW fans be tempted into greater engagement through a politics that centers moralism and the policing of every petty interaction. We need a material analysis, buttressed with a sense of humor and a recognition of human fallibility, that connects the fight for a better world to the immediate interests of the majority of the population.

The point of such an analysis isn't to tell people not to worry about the sort of spiritual and existential concerns that drive young men into the arms of Peterson. On the contrary, the impulse behind left-wing politics is the desire to create a world in which our lives aren't dominated by economic concerns— where people have the free time and energy to explore, to meditate, to read novels (or try their hand at writing them), and to pursue meaningful relationships because they're free from the workplace tyranny that leaves them too exhausted at the end of every day to do anything but watch Netflix, mindlessly skim through their social media feeds, or, God forbid, game (and that, for many, is the *best case scenario*). The central contradiction in Peterson's message is that he both uncritically celebrates capitalist "free markets" and sounds the alarm at the destructive toll those markets inevitably take on relationships and communities. Our message, however, must reject Peterson's traditionalist and pseudo-libertarian worldview in favor of a vision in which everyone has the economic freedom—as in, freedom from economics—to pursue their own preferred vision of the good life. Want to have a traditional family and take your seven kids to a traditional church? Go for it. Want to live in a tri-sexual compound and practice Wiccanism? Do that. Where Peterson wants to "enforce" monogamy, socialists like me want to give everyone the freedom to make more meaningful choices in their lives by creating a world in which financial stress doesn't make it difficult to maintain relationships, people who want to start families can, and we aren't all too overworked, over-stressed, and socially atomized to go out and meet people in the first place (or too afraid of indigency that we stay in toxic relationships).

Even in the most tyrannical institution of capitalist societies— namely, the workplace—social democratic reforms expand the sphere of human freedom. Think about the way that many people are so afraid of losing their family's health insurance that they do whatever they can to please their bosses, whether this means working overtime without pay or suffering workplace harassment. Something as simple as Medicare for All, which has been on the books for decades in Canada—where it's so popular that even conservative politicians have to at least pretend to support it—can do a lot to alleviate this unfreedom. Similar considerations apply to programs ranging from tuition-free public college to state-sponsored childcare schemes to reducing the workweek to 3 or 4 days.

The recent reemergence of social movements around the world, from Haiti to Lebanon, Chile to Sudan and the noble effort of Jeremy Corbyn and the powerful force and potential election of Bernie Sanders to the presidency, attests to the widespread appeal of this vision. It's important to emphasize, though, that while social democracy is an immensely valuable step in the right direction—especially in a world ravaged by decades of neoliberalism and US-led militarism—for two reasons it ultimately won't be enough.

The first reason is ideological. Why should anyone accept an economic order in which a minority of the population has the resources to own businesses and everyone else has to submit to their authority for (at least) eight out of every 16 waking hours? (Yeah, yeah, some people are upwardly mobile, but there are only so many lifeboats to save the working class. We can't all be athletes, business prodigies, or even podcasters and YouTube talk show hosts.) And contrary to the delusions of the #YangGang, the combination of robots and a Universal Basic Income isn't going to result in any kind of desirable alternative. Even if things *did* play out that way, which they won't, that's just a recipe for ever-greater division between rich and poor. As Ben Burgis said when we discussed Yang on TMBS, the #YangGang's dystopia might look a lot like the Roman Empire, in which wealthy aristocrats monopolized farmable land and forced the poor to flock to Rome to live on a miserly grain ration. Or as I say, worry about capitalism, not robots.

The Mondragon Corporation in Spain showcases at least some aspects of what a better alternative might look like. Another great example is Cooperation Jackson, a cooperative based in Jackson, Mississippi, which grew out of the pioneering leadership of Chokwe Lumumba and the New Africa Movement. One of the latter's many victories was to elect Chokwe Lumumba and then, after his untimely death in 2014, his son Chokwe Antar Lumumba, mayor of Jackson. In the words of Malaika Jabali, who discussed Cooperation Jackson with me on TMBS, Chokwe Antar Lumumba is "the most progressive mayor in America." Cooperation Jackson's initiatives from the grassroots and in local government have ranged from investment in community-supported agriculture to worker-owned businesses and citizen participation in setting municipal budgets.

No one knows for certain what it might look like to apply

similar principles to the construction of a new national—and eventually international—economic order. However, a combination of nationalizations of big banks and other "commanding heights" of the economy and the promotion of worker-run enterprises over traditional hierarchical businesses would be key ingredients in the initial phases of any transition away from capitalism. I've been lucky enough to talk to the great Marxist economist Richard Wolff about these issues many times on TMBS, and I would strongly recommend his book *Democracy at Work: A Cure for Capitalism.*

When it comes to automation, the crucial issue ignored by the #YangGang is the question of ownership. Even if all their fantasies of everything being automated come true, if the machines were privately owned, they would enrich the owners while the rest of us try to survive on the financial ration of a UBI check. In contrast, if the machines are collectively owned, then automation means we can all decide to work fewer hours so we have more time for everything that matters in life and fulfill our highest human potential. As Marx said, in an ideal world humans would "hunt in the morning, fish in the afternoon, rear cattle in the evening, criticize after dinner...without ever becoming a hunter, fisherman, herdsman, or critic." *That's* a vision worth fighting for.

The second, more pragmatic reason to go beyond social democracy to more radically socialist proposals is that at a certain point we'll need to start democratizing the economy to defend the gains we've already made. As long as the capitalist class retains its economic power, owners will always try to roll back workers' advances. The history of the twentieth century shows this process playing out over and over again in all areas of the world, from the Global North to the Global South.

Thus far I've been talking mostly about economics, but there's a distinction worth making between XXX and YYY. Some with a material politics who rightly reject the emotional toxicity and shallow analyses of the woke left have fallen into a credulous conservativism that all too often validates the woke brigades' worst suspicions and indeed bolsters genuinely bad ideas. It's entirely possible to roll your eyes at denunciations of "problematic" comedians and center class in your political analysis while fighting with all your heart and soul against the Trump administration putting immigrant children in concentration camps. If you find yourself reacting to both the policing of comedy and the protests against serious human rights abuses at the southern border as if they were equally unserious liberal preoccupations, you've jettisoned your sense of perspective and lost touch with important left principles—not to mention your basic humanity.

What we need is a cosmopolitan socialism premised on real material needs that expresses itself in criticism, art, movement-building, and anything else that drives politics. Again, following Gramsci, we need an integral approach that fuses universal desires, aspirations, and material concerns with a recognition that we do in fact live in a globalized, interconnected, and neoliberal world still defined by grotesque inequality, ecological crisis, and resurgent right-wing authoritarianism. Our approach can't simply be to tell people they're wrong to be concerned with the cultural issues that define much of their lives or to dismiss the importance of oppression that doesn't always take an economic form. Rather, we need to recapture the spirit that appealed to young Bengali Marxist MN Roy in 1920. Bhaskar Sunkara, the founding editor of *Jacobin*, quotes Roy in his recent *The Socialist Manifesto*.

To MN Roy, the Communist International's Second Congress was a revelation. "For the first time," he remarked, "brown and yellow men met with white men who were not overbearing imperialists but friends and comrades."

The point isn't to valorize the mixed record of the capital-C Communist movement on issues of racial and national liberation.

Though communist parties *have* in certain instances contributed to racial and anticolonial justice, especially in Cuba, Vietnam, South Africa, and the United States, in the Soviet Union and the People's Republic of China authoritarian communists suppressed national, ethnic, and religious minorities in indefensible ways. Rather, the point is to highlight that the type of comradeship and solidarity across racial and national lines that Roy wrote about is going to have to be central to any kind of viable movement to achieve a better world; indeed, it's absolutely central to the democratic and humanistic Marxism with which I identify.

The need to build transracial and transnational solidarities is important for both strategic and moral reasons. When it comes to the struggle to enact even relatively modest social democratic gains in the United States—what the historian and activist Harvey Kaye calls *completing the New Deal*—the electoral math makes the strategic calculation straightforward. Bernie Sanders won the Michigan primary in 2016 because he won the Arab-American vote in Dearborn. Going back to the 2016 general election, Hillary Clinton lost—despite winning the national popular vote—because traditional Democratic constituencies in places like Detroit, Flint, and Milwaukee weren't inspired by her message and didn't come out to vote. Granted, the people of Michigan voted for Bernie because his economic message inspired in a way Hillary's centrism never did; unsurprisingly, black and Arab-American voters in the Upper Midwest have a lot of the same bread-and-butter concerns as their white counterparts. (The media's habit of treating those economic concerns as an exclusively "white" issue is maddening given that by any measure the effects of deindustrialization have come down hardest on the rustbelt's black population. See Malaika Jabali's work on Wisconsin in 2016.) Nonetheless, it's hard not to imagine that Clinton's repellent history on issues like the mass incarceration of "superpredators" didn't play a role in adding to the disgust and indifference that resulted in non-white voter turnout being much lower than it was in 2012, or that at least a few voters in Dearborn didn't care that Sanders was less likely than Clinton to kill their relatives in future drone strikes. In other words, the future of the US left depends on making common cause with people of all colors.

This is likewise true on the international level. Global migration patterns have created surprisingly diverse populations, even in European countries that Americans often think of as exclusively white. And moreover, even in Western nations that really are monochromatic, workers' interests are inexorably tied to those of their counterparts in the Global South. This is because the threat of capital flight to lower-wage countries threatens social democratic advances and undermines workers' bargaining power on the shop floor. The last several decades of labor history have demonstrated that worker attempts to transcend capitalism are often blocked by anti-labor adjustments that nationalized or cooperatized industries make to compete with exploitative private firms in a global marketplace. Furthermore, there is also the fact that historically the techniques of oppression that police and other elites use to prevent worker activism are always perfected first in the Global South. Thus, when Italian unionists refused to load a Saudi ship with materials for the emirates' genocidal campaign in Yemen, they were acting in accordance with very real strategic imperatives. Given Italy's long history of socialist and communist organizing, it's possible that this was even a conscious motivation.

This reminds me of a story told by the late socialist and war resistor David McReynolds, who was for decades a member of the now-defunct Socialist Party of America (SPA). The SPA had a long and proud history dating back to Eugene V. Debs in the early twentieth century. In its final decades, though, the party sometimes erred on the side of an exaggerated concern about Stalinism that led some of its leaders to downplay or ignore the far more real threat of American imperialism. This culminated in a split over the war in Vietnam, with some members supporting the intervention and some rejecting it. McReynolds—who was one of the very first people to publicly burn his draft card in 1965—was on the right side of that split, just as he landed on the right side of just about every other major debate of the late twentieth century.

This story takes place in 1957, 8 years before McReynolds burned his draft card. That year, the Soviet Union launched the Sputnik I satellite, which was the first human-made satellite to orbit Earth. McReynolds was watching coverage of the launch on a television in a bar in Greenwich Village and in his excitement ran outside and started a conversation with the first person he saw:

"We did it!"

"What?"

"Sputnik! We put something into the sky that didn't come down!"

"You mean the Russians did."

"No. We. Us. The human race."

Some leftists might resist the adoption of this kind of universal perspective—*we, us, the human race*—for at least two reasons. On one hand, Marxists might worry that it erases the distinction between different groups of humans with divergent economic interests. To my mind, though, this ignores the humanistic impulse that leads one to become a socialist in the first place. While it's true that struggling for a democratized economy means struggling *against* the segment of the population that benefits from the current undemocratic order, it's also true that, as Engels put it in *The Anti-Dühring*, ending the division of society into contending social classes creates the possibility for "a really human morality" universal in character.

On the other hand, some might be anxious that, though

this kind of humanism claims to be universal, people who speak in these terms often have a perspective that is painfully particularistic and, specifically, "Western." As the economist Amartya Sen emphasizes, if freedom and equality and solidarity and the rest are genuinely universal human values—and like Sen, I believe they are—we should be able to root them in a multiplicity of cultural traditions. The kind of global socialist internationalism I'm talking about doesn't mean that everyone has to do all of their thinking and theorizing in English—or even Esperanto. It means building a truly global intellectual and political culture with roots in a diversity of societies. Put another way, just because the messengers of universal humanism aren't necessarily who we want them to be, doesn't mean that the message itself isn't just.

Sen, for instance, has noted the importance of the concept of "freedom" in the Buddhist tradition. In an address to the Carnegie Council, he complicated the standard-issue view of Confucianism as quiescent, quoting the classical Chinese philosopher as saying that one must tell "the truth even if it offends." As Sen put it, "Those in charge of censorship in Singapore or Beijing would take a very different view. Confucius is not averse to practical caution and tact, but does not forgo the recommendation to oppose a bad government." Shifting to India, Sen centers the role of argument in the Indian tradition. He quotes the Bengali poet Ram Mohan Ray, who wrote:

Just imagine how terrible it will be on the day you die,
Others will go on speaking, but you will not be able to respond.

Sen also highlights the third century BC Emperor Ashoka, who "promoted public ethics after being horrified by the carnage he saw in his own victorious battle against the king of Kalinga (now Orissa)." In addition to converting to Buddhism and sending missionaries to promote the religion abroad, Ashoka "covered the country" with stone inscriptions inscribed with these and similar sayings:

A man must not do reverence to his own sect or disparage that of another man without reason. Depreciation should be for specific reason only, because the sects of other people all deserve reverence for one reason or another.

By thus acting, a man exalts his own sect, and at the same time does service to the sects of other people. By acting contrawise, a man hurts his own sect, and does disservice to the sects of other people. For he who does reverence to his own sect while disparaging the sects of others wholly from attachment to his own, with intent to enhance the splendor of his own sect, in reality by such conduct inflicts the severest injury on his own sect.

Anyone harboring illusions that tolerance and pluralism are uniquely and characteristically "Western" values might want to reflect on the fact that Ashoka commissioned that inscription centuries *before* his fellow emperor Caligula entertained himself and the rest of Rome by burning early Christians alive. Good and liberating ideas, like bad and reactionary ones, have thrived in a variety of cultural settings.

Consider the Trinidadian Marxist CLR James, who is most famous for his classic history of the Haitian Revolution, *The Black Jacobins*. (The cover art from that book, by the way, was a pretty clear inspiration for iconography used by a magazine with a similar name founded by Bhaskar Sunkara, who is himself from Trinidad.) Anti-colonialism was one of James' core commitments. Despite (or maybe not "despite") that, as Ralph Lennard discusses in his fascinating article "CLR James Rejected the Posturing of Identity Politics," James was never simply "anti-Western."

For James, the emancipation of the black mind would come from embracing the works of "dead white men" such as Socrates, Sophocles, Cervantes, Shakespeare, Thackeray, and Dickens, as much as the works of WEB Du Bois, Aime Cesairé, Richard Wright, Ralph Ellison, and Toni Morrison.

He believed the European canon provided black people with a means of empowering themselves both culturally and intellectually, to broaden their imagination, to liberate them from the tyranny of geography and race, and help them transcend their particularity and enter into a universal conversation across color lines, based on a shared humanity.

Leonard quotes an exchange in which James was asked, "Should Shakespeare and Rembrandt and Beethoven matter to Caribbean people?" James said he didn't like the question. He granted that if it meant that "Caribbean writers should be aware that there are emphases in their writing that we owe to non-European, non-Shakespearean roots, and the past in music which is not Beethoven," he had no objection, but in the final analysis "the Caribbean people are people," and as great artists Shakespeare and Rembrandt and Beethoven should matter to all people regardless of their racial, ethnic, gender, or class identity.

Of course, and as Leonard observed in a conversation with me, much of the "ancient Western tradition" was in fact highly geographically and intellectually diverse and included African and pan-Asian sources that are misleadingly remembered—and misleadingly whitened—as merely "Greek" or "Roman." But to underline the larger point I am trying to make: Instead of policing each other's influences and enjoyments for evidence of "cultural appropriation," we should all strive to emulate the curiosity and rigor of the great Christian revolutionary intellectual Cornel West, who explores the echoes between Anton Chekhov and the blues with no interest in drawing artificial lines between cultures.

In making this point, I'm not claiming that there is nothing wrong with some of the things people have labeled "cultural appropriation." Jay-Z was absolutely right to get "what they did to the Cold Crush"—and to get himself on the cover of *Fortune* Magazine. I agree with what frequent TMBS guest and crew member Brihanna Joy Gray—who is now Bernie Sanders' press secretary—said in a brilliant 2017 article for *Current Affairs*:

> If the definition of "appropriation" had stayed narrow, it would be easy enough to defend. It's obvious it's insulting and upsetting for a white person to casually sport a feathered headdress. They're items of deep symbolic meaning to the people who originated them, bestowed in recognition of great achievements. Treating them as party hats cheapens and dishonors them...but as the "appropriation" concept has been used to object to many formally innocuous forms of cultural mixing, certain criticisms of the term become increasingly credible. The more things are stuffed under the "cultural appropriation tent," the more legitimate the concern that it may put limits on creativity, cultural exchange, and innovation.

To this list of concerns, I would add: Remaining forever fearful of cultural appropriation will smother the international socialist project I envision by contributing to what Adolph Reed has disparagingly termed "essentialism," the almost metaphysical belief that culture stands apart from politics and economics. Simply put, there is no magical or particular essence that gives people born into a culture the right to deny those who are not from that culture access to art, ideas, music, and the like. We should all be open to all cultures and should in fact embrace and encourage cultural exchange and syncretism.

Similarly, when it comes to political strategy, we should draw inspiration from the history of transnational organizing, in which people from various cultures worked together to overcome some of the twentieth century's worst oppressions. Take the history of South Africa's African National Congress (ANC), the party of Nelson Mandela. Founded in 1912 as the South African Native National Congress, the ANC helped smash apartheid and, in fact, still governs the country today. The party's history exemplifies some of the absolute best of the international socialist tradition—and, tragically, also shows how neoliberalism, appealing to narrowly-conceived identity politics, and authoritarianism can undermine revolutionary momentum. But I'll get to that part in a moment. First, let's talk about the ANC's incredible synthesis.

In 1955, the ANC's Freedom Charter articulated the goals and aspirations of a liberated South Africa in a strikingly international socialist idiom:

1: We, the People of South Africa, declare for all our country and the world to know: that South Africa belongs to all who live in it, black and white, and that no government can justly claim authority unless it is based on the will of all the people; that our people have been robbed of their birthright to land, liberty and peace by a form of government founded on injustice and inequality; that our country will never be prosperous or free until all our people live in brotherhood, enjoying equal rights and opportunities; that only a democratic state, based on the will of all the people, can secure to all their birthright without distinction of color, race, sex or belief;

2: All people shall have equal right to use their own languages, and to develop their own folk culture and customs;

3: The national wealth of our country, the heritage of all South Africans, shall be restored to the people; The mineral wealth beneath the soil, the banks and monopoly industry shall be transferred to the ownership of the people as a whole;

4: Restrictions of land ownership on a racial basis shall be ended, and all the land redivided amongst those who work it, to banish famine and land hunger; The state shall help the peasants with implements, seed, tractors and dams to save the soil and assist the tillers; Freedom of movement shall be guaranteed to all who work on the land.

The ANC would go on to form its armed wing—the Umkhonto we Sizwe (UWS, "the spear of the nation") in partnership with the South African Communist Party (SACP). The UWS's founding was itself internationalist; members included Mandela as well as Jewish Communist revolutionary Ronnie Kasrils, who after apartheid would serve in the cabinets of Mandela and Thabo Mbeki. Indeed, the ANC framed its ferocious opposition to apartheid and white supremacy in terms of a commitment to "non-racialism." Over the course of decades of struggle, it held together an incredible international coalition that included educated middle-class black professionals, communists, trade unionists, and democracy campaigners. And the party was aided by people the world over. To the Soviet Union's credit, it backed the ANC, as did Olaf Palme's social democratic Sweden. Within and across cultures and within and across borders, the ANC built a lasting transnational coalition that helped the party end apartheid in the 1990s.

Much of the Freedom Charter was incorporated into the constitution of liberated South Africa, making it one of the most genuinely progressive constitutions the world has ever seen. And though Mandela's notion of a "rainbow nation" might sound like something plucked from the sanitized world of a Disney movie, it has roots in a noble revolutionary tradition to which Mandela stayed true for his life's duration, even as he made uncomfortable compromises to secure South Africa's liberation in a post-Cold War, and increasingly neoliberal, world order. For instance, Mandela was steadfast in his support for Cuba; remained a ferocious advocate for Palestinian rights; and responded with moral indignation to the US invasion of Iraq.

There is a tremendous amount to learn from the history and theory of the ANC. To be sure, the record of the ANC in power doesn't always match the best of its history during the decades of struggle. Simply put, the ANC has been unable to overcome the "economic apartheid" that continues to limit its, and South Africa's, prospects. As Ronnie Kasril, who I've had the honor of interviewing several times on TMBS and who is now a dissident, has argued, the ANC achieved political liberation at the expense of the economic liberation promised in the Freedom Charter. The unfulfilled promise of this next phase of liberation has kept the obscene economic inequality that has long characterized South Africa in place; in turn, many of the old structures of racial hierarchy remain, and the country continues to generate political forces that traffic in demagoguery, crony capitalism, and a shallow technocratic liberalism.

These problems are hardly unique to South Africa. Much of what has gone wrong there are local manifestations of global trends. What the world needs to counter these forces is a renewed commitment to the kind of international socialist politics embodied in the Freedom Charter. This book has been my contribution to that cause.

CULTURE, SOCIETY & POLITICS

The modern world is at an impasse. Disasters scroll across our smartphone screens and we're invited to like, follow or upvote, but critical thinking is harder and harder to find. Rather than connecting us in common struggle and debate, the internet has sped up and deepened a long-standing process of alienation and atomization. Zer0 Books wants to work against this trend. With critical theory as our jumping off point, we aim to publish books that make our readers uncomfortable. We want to move beyond received opinions.

Zer0 Books is on the left and wants to reinvent the left. We are sick of the injustice, the suffering and the stupidity that defines both our political and cultural world, and we aim to find a new foundation for a new struggle.

If this book has helped you to clarify an idea, solve a problem or extend your knowledge, you may want to check out our online content as well. Look for Zer0 Books: Advancing Conversations in the iTunes directory and for our Zer0 Books YouTube channel.

Popular videos include:

Žižek and the Double Blackmain

The Intellectual Dark Web is a Bad Sign

Can there be an Anti-SJW Left?

Answering Jordan Peterson on Marxism

Follow us on Facebook
at https://www.facebook.com/ZeroBooks and Twitter at https://
twitter.com/Zer0Books

Bestsellers from Zer0 Books include:

Give Them An Argument
Logic for the Left
Ben Burgis
Many serious leftists have learned to distrust talk of logic. This is
a serious mistake.
Paperback: 978-1-78904-210-8 ebook: 978-1-78904-211-5

Poor but Sexy
Culture Clashes in Europe East and West
Agata Pyzik
How the East stayed East and the West stayed West.
Paperback: 978-1-78099-394-2 ebook: 978-1-78099-395-9

An Anthropology of Nothing in Particular
Martin Demant Frederiksen
A journey into the social lives of meaninglessness.
Paperback: 978-1-78535-699-5 ebook: 978-1-78535-700-8

In the Dust of This Planet
Horror of Philosophy vol. 1
Eugene Thacker
In the first of a series of three books on the Horror of Philosophy,
In the Dust of This Planet offers the genre of horror as a way of
thinking about the unthinkable.
Paperback: 978-1-84694-676-9 ebook: 978-1-78099-010-1

The End of Oulipo?
An Attempt to Exhaust a Movement
Lauren Elkin, Veronica Esposito
Paperback: 978-1-78099-655-4 ebook: 978-1-78099-656-1

Capitalist Realism
Is There no Alternative?
Mark Fisher
An analysis of the ways in which capitalism has presented itself
as the only realistic political-economic system.
Paperback: 978-1-84694-317-1 ebook: 978-1-78099-734-6

Rebel Rebel
Chris O'Leary
David Bowie: every single song. Everything you want to know,
everything you didn't know.
Paperback: 978-1-78099-244-0 ebook: 978-1-78099-713-1

Kill All Normies
Angela Nagle
Online culture wars from 4chan and Tumblr to Trump.
Paperback: 978-1- 78535-543-1 ebook: 978-1-78535-544-8

Cartographies of the Absolute
Alberto Toscano, Jeff Kinkle
An aesthetics of the economy for the twenty-first century.
Paperback: 978-1-78099-275-4 ebook: 978-1-78279-973-3

Malign Velocities
Accelerationism and Capitalism
Benjamin Noys
Long listed for the Bread and Roses Prize 2015, *Malign Velocities*
argues against the need for speed, tracking acceleration
as the symptom of the ongoing crises of capitalism.
Paperback: 978-1-78279-300-7 ebook: 978-1-78279-299-4

Meat Market
Female Flesh under Capitalism
Laurie Penny
A feminist dissection of women's bodies as the fleshy fulcrum of
capitalist cannibalism, whereby women are both consumers and
consumed.
Paperback: 978-1-84694-521-2 ebook: 978-1-84694-782-7

Babbling Corpse
Vaporwave and the Commodification of Ghosts
Grafton Tanner
Paperback: 978-1-78279-759-3 ebook: 978-1-78279-760-9

New Work New Culture
Work we want and a culture that strengthens us
Frithjoff Bergmann
A serious alternative for mankind and the planet.
Paperback: 978-1-78904-064-7 ebook: 978-1-78904-065-4

Romeo and Juliet in Palestine
Teaching Under Occupation
Tom Sperlinger
Life in the West Bank, the nature of pedagogy and the role of a
university under occupation.
Paperback: 978-1-78279-637-4 ebook: 978-1-78279-636-7

Ghosts of My Life
Writings on Depression, Hauntology and Lost Futures
Mark Fisher
Paperback: 978-1-78099-226-6 ebook: 978-1-78279-624-4

Sweetening the Pill
or How We Got Hooked on Hormonal Birth Control
Holly Grigg-Spall
Has contraception liberated or oppressed women?
Sweetening the Pill breaks the silence on the dark side of hormonal
contraception.
Paperback: 978-1-78099-607-3 ebook: 978-1-78099-608-0

Why Are We The Good Guys?
Reclaiming your Mind from the Delusions of Propaganda
David Cromwell
A provocative challenge to the standard ideology that Western
power is a benevolent force in the world.
Paperback: 978-1-78099-365-2 ebook: 978-1-78099-366-9

The Writing on the Wall
On the Decomposition of Capitalism and its Critics
Anselm Jappe, Alastair Hemmens
A new approach to the meaning of social emancipation.
Paperback: 978-1-78535-581-3 ebook: 978-1-78535-582-0

Neglected or Misunderstood
The Radical Feminism of Shulamith Firestone
Victoria Margree
An interrogation of issues surrounding gender, biology, sexuality, work and technology, and the ways in which our imaginations continue to be in thrall to ideologies of maternity and the nuclear family.
Paperback: 978-1-78535-539-4 ebook: 978-1-78535-540-0

How to Dismantle the NHS in 10 Easy Steps (Second Edition)
Youssef El-Gingihy
The story of how your NHS was sold off and why you will have to buy private health insurance soon. A new expanded second edition with chapters on junior doctors' strikes and government blueprints for US-style healthcare.
Paperback: 978-1-78904-178-1 ebook: 978-1-78904-179-8

Digesting Recipes
The Art of Culinary Notation
Susannah Worth
A recipe is an instruction, the imperative tone of the expert, but this constraint can offer its own kind of potential. A recipe need not be a domestic trap but might instead offer escape – something to fantasise about or aspire to.
Paperback: 978-1-78279-860-6 ebook: 978-1-78279-859-0

Most titles are published in paperback and as an ebook. Paperbacks are available in traditional bookshops. Both print and ebook formats are available online.
Follow us on Facebook
at https://www.facebook.com/ZeroBooks
and Twitter at https://twitter.com/Zer0Books